INTERNATIONAL CENTRE FOR MECHANICAL SCIENCES

COURSES AND LECTURES No. 86

SILVIU GUIASU

UNIVERSITY OF BUCHAREST

MATHEMATICAL STRUCTURE OF FINITE RANDOM CYBERNETIC SYSTEMS

LECTURES HELD AT THE DEPARTMENT
FOR AUTOMATION AND INFORMATION
JULY 1971

UDINE 1971

SPRINGER-VERLAG WIEN GMBH

Originally published by Springer-Verlag Wien-New York in 1972

ISBN 978-3-211-81174-0 ISBN 978-3-7091-2802-2 (eBook)

DOI 10.1007/978-3-7091-2802-2

P R E F A C E

The material contained in these lecture notes was covered by the author during the course held at the INTERNATIONAL CENTRE FOR MECHANICAL SCIENCES UDINE; in June-July 1971.

I am grateful to all the Authorities of this magnificent Centre for giving me the opportunity of delivering the course and especially to Professor L. SOBRERO.

I am indebted to Professor GIUSEPPE LONGO for his kind observations, and useful advice. His charm and our discussions will remain unforgettable.

I was also impressed by the high level of technical assistance supplied by the International Centre for Mechanical Sciences.

Udine, July 1971

S. Guiașu

Introduction

Informational models given for learning systems ([9], [12]), for random automata ([10], [12], [15]) and for systems with strategies ([11], [12]) can be managed naturally. In spite of seeming diversity, there is a common mathematical structure for all these cybernetic systems. This mathematical structure proper to a large group of cybernetic systems with random behaviour will be discussed fully in the following pages.

Formally, this mathematical structure does not exceed the category theory and on the other hand the application of the instruments of the category theory to the study of some deterministic cybernetic systems (essentially for deterministic automata only) is no longer a novelty (see [1],[2] [6]). Nevertheless, even if the finite cybernetic systems with random behaviour are categories too, it is easy to see that almost all results of category theory are not directly applicable to the study of these systems. The reason for this strange situation lies in the fact that the morphisms are habitually functions in almost all usual applications of the category theory (and in the applications to the deterministic cybernetic systems too) whereas in the category of finite cybernetic systems with random behaviour, these morphisms are essentially random correspondences given by transition matrices, i.e. by stochastic matrices. On the other hand, these random cybernetic systems induce many specific problems which do not appear in the usual category theory. The problem of the replacement of the random morphism by an ε deterministic morphism is a case in point, being the core of the codifications in cybernetic systems. It concerns random correspondence given by a stochastic matrix, almost deterministic correspondence i.e. a correspondence given by a usual function with error smaller than ε . Another specific problem of the study of the processes describes the time-evolution of the finite random cybernetic systems. Such a process is composed of a family of morphisms characterizing the whole evolution of these cybernetic systems. The employmentof the diagram techniques and of the instruments of information theory will be fruitful in the whole categorial approach of finite random cybernetic systems.

The contents of the paper can be summarized as follows. The first chapter contains the definition of the finite random categories, abbreviated by FR-categories, including the main classes of morphisms and their techniques which will be utilized in the paper. The second chapter includes examples of FR-categories, namely the noisy communication systems without memory, the finite random automata, the two-person games and the learning systems. Frequent reference will be made to these examples throughout the paper in order to apply the general theory for arbitrary FR-categories, whose processes will be studied in Chapter three. Such a process in a given FR-category contains those morphisms which characterize completely the whole time-evolution of the random cybernetic system described by the respective FR-category. In the same chapter the general theory is applied both to the learning process from an arbitrary learning system, and to the process describing the evolution from the random automata as well as to the process occurring in an arbitrary two-person game. The fourth chapter, which together with chapter three is one of the largest sections of the paper, and considers the problem of the reduction of one arbitrary random morphism to an ε-deterministic morphism containing applications of the general theory to the codification in Markovian communication systems and to the codification in random automata. Applications of the most rational algorithm of recognition to the codind and decoding problem are also given in this chapter together with some observations about the algebraic and probabilistic theory of codes.

Chapter 1
DEFINITION OF FINITE RANDOM CATEGORIES

To give a category $\mathcal{C}$ it means to give:

a) A set $Ob(\mathcal{C})$ whose elements are called the objects of $\mathcal{C}$;

b) For every pair of objects $X,Y \in Ob(\mathcal{C})$ a set $Hom_{\mathcal{C}}(X,Y)$ (or simply $Hom(X,Y)$) whose elements are called morphisms (or arrows) from X to Y i.e. with the source X and ending Y. In denoting the source and the ending of an arbitrary morphism u by $S(u)$ and respectively $E(u)$, given that $u \in Hom_{\mathcal{C}}(X,Y)$ one gets $S(u)=X$ and $E(u)=Y$. The sets $Hom_{\mathcal{C}}(X,Y)$ are mutually disjoint, i.e. every morphism has a single source and a single ending;

c) For every three objects $X,Y,Z \in Ob(\mathcal{C})$ an application

$$Hom_{\mathcal{C}}(X,Y) \times Hom_{\mathcal{C}}(Y,Z) \longrightarrow Hom_{\mathcal{C}}(X,Z)$$

called the <u>composition</u> of the morphisms which associates with every pair of morphisms $u \in Hom_{\mathcal{C}}(X,Y)$, $v \in Hom_{\mathcal{C}}(Y,Z)$ one morphism from $Hom_{\mathcal{C}}(X,Z)$ being denoted by $v_0 u$ or vu.

Given that category $\mathcal{C}$ is <u>associative</u> the composition of the morphisms is associative,

$$w_0(w_0 u) = (w_0 v)_0 u$$

no matter what the morphisms

$$u \in Hom_{\mathcal{C}}(X,Y)\ , \quad v \in Hom_{\mathcal{C}}(Y,Z)\ , \quad w \in Hom_{\mathcal{C}}(Z,U)\ .$$

Let us suppose Category $\mathcal{C}$ to be a category with identical morphisms if for every object $X \in Ob(\mathcal{C})$ there exists a morphism $1_X \in Hom_{\mathcal{C}}(X,X)$ called the identical morphism of the object X, or the identity of X, so that $1_X \circ u = u$ and $v \circ 1_X = v$ for every morphism u with the ending X and every morphism v with the source X.

We shall denote an arbitrary morphism $u \in Hom_{\mathcal{C}}(X,Y)$ by $u:X \longrightarrow Y$ or frequently by

$$X \xrightarrow{u} Y\ .$$

We shall also use $\mathcal{M}(\mathcal{C})$ to denote the set of all morphisms of the category $\mathcal{C}$, i.e.

$$\mathcal{M}(\mathcal{C}) = \cup Hom_{\mathcal{C}}(X,Y)$$

where the union is taken over all objects X,Y from $Ob(\mathcal{C})$. Obviously, in a category with identical morphisms there is one-to-one correspondence between the objects of the category $\mathcal{C}$ and the identical morphisms $X \longleftrightarrow 1_X$.

If $\mathcal{C}$ is a category, let the dual category of $\mathcal{C}$ be denoted by $\mathcal{C}^0$, defined as follows such that

a) $Ob(\mathcal{C}^0) = Ob(\mathcal{C})$;

b) $Hom_{\mathcal{C}^0}(X,Y) = Hom_{\mathcal{C}}(Y,X)$

whatever be the objects X, Y;

c) the composition of v and u in $\mathcal{C}^0$ is equal to the composition of u and v in $\mathcal{C}$.

A category $\mathcal{C}'$ is called a <u>subcategory</u> of $\mathcal{C}$ if

a) $Ob(\mathcal{C}') \subset Ob(\mathcal{C})$;

b) $Hom_{\mathcal{C}'}(X,Y) \subset Hom_{\mathcal{C}}(X,Y)$

for every pair of objects X,Y from $Ob(\mathcal{C}')$;

c) The composition of the morphisms in $\mathcal{C}'$ is indicated, is induced, by the composition of the morphisms in $\mathcal{C}$.

A subcategory $\mathcal{C}'$ of $\mathcal{C}$ is called a <u>full</u> subcategory if

$$Hom_{\mathcal{C}'}(X,Y) = Hom_{\mathcal{C}}(X,Y)$$

whichever be the pair of objects X,Y from $Ob(\mathcal{C}')$.

A subcategory $\mathcal{C}'$ of $\mathcal{C}$ is called a <u>rich</u> category if

$$Ob(\mathcal{C}') = Ob(\mathcal{C}).$$

Obviously, a subcategory $\mathcal{C}'$ of the category $\mathcal{C}$ which is at the same time both rich and full will coincide with $\mathcal{C}$.

We shall say that a category is an <u>FR-category of Σ-type</u> (finite random category of Σ-type) if:

a) The objects are finite sets;

b) The morphisms are stochastic matrices, i.e. no matter what the objects X,Y if $\mathrm{Hom}(X,Y) \neq \emptyset$ then an arbitrary morphism $u \in \mathrm{Hom}(X,Y)$ is given by

$$u = (p(y|x))_{\substack{x\in X \\ y\in Y}}$$

where

$$p(y|x) \geq 0 \text{ whichever be } x\in X\,,\ y\in Y\,,$$

$$\sum_{y\in Y} p(y|x) = 1 \text{ for every } x\in X\,. \tag{1.1}$$

(It is possible to exist objects X,Y such that $\mathrm{Hom}(X,Y)=\emptyset$.) The term "random morphism" will henceforth be used to describe this kind of morphism, which will, in future be denoted by

$$p(y|x) : X \longrightarrow Y$$

or by

$$X \xrightarrow{\ p(y|x)\ } Y$$

understanding that x belongs to the set X and y belongs to the set Y. Although the term $p(y|x)$ will be used throughout the paper to denote a random morphism, it will, nevertheless be regarded as representing the stochastic matrix

$$(p(y|x))_{\substack{x \in X \\ y \in Y}} \,.$$

c) For every system of three objects X,Y,Z we define the composition of two arbitrary morphisms

$$p(y|x) \in \mathrm{Hom}(X,Y) \quad , \quad p(z|y) \in \mathrm{Hom}(Y,Z)$$

to be

(1.2) $$X \xrightarrow{p(y|x)} Y \xrightarrow{p(z|y)} Z \;=\; X \xrightarrow{p(z|x)} Z$$

where

(1.3) $$p(z|x) = \sum_{y \in Y} p(z|y)p(y|x)$$

no matter what $x \in X, z \in Z$ which will be denoted as follows

(1.4) $$p(z|x) = p(z|y) \circ p(y|x) \,.$$

d) For every object X there exists the identical morphism $1_X \in \mathrm{Hom}(X,X)$ defined by

$$1_X = (p(x'|x))_{\substack{x \in X \\ x' \in X}}$$

(1.5) $$p(x'|x) = \delta_{x',x} = \begin{cases} 1 & \text{if } x' = x \\ 0 & \text{if } x' \neq x \,. \end{cases}$$

Let us now show that an FR-category of $\sum$-type defined aboveis indeed an associative category with identical morphisms.

In this direction it is necessary to prove several propositions.

PROPOSITION 1.1: *The composition of the random morphisms given by the equality (1.4) is a random morphism too, namely*

$$p(z|y)\circ p(y|x)\in \mathrm{Hom}(X,Z) .$$

PROOF: From (1.3) and taking into account that $p(y|x)$ and $p(z|y)$ are both random morphisms we have

$$p(z|x) = p(z|y)\circ p(y|x) \geqslant 0$$

for every $x\in X, z\in Z$ and also

$$\sum_{z\in Z} p(z|x) = \sum_{z\in Z}\sum_{y\in Y} p(z|y)\,p(y|x) =$$

$$= \sum_{y\in Y}\Big(\sum_{z\in Z} p(z|y)\Big)p(y|x) = \sum_{y\in Y} p(y|x) = 1 .$$

q.e.d.

The equality (1.2) of the random morphisms will be written both in the form of the diagram

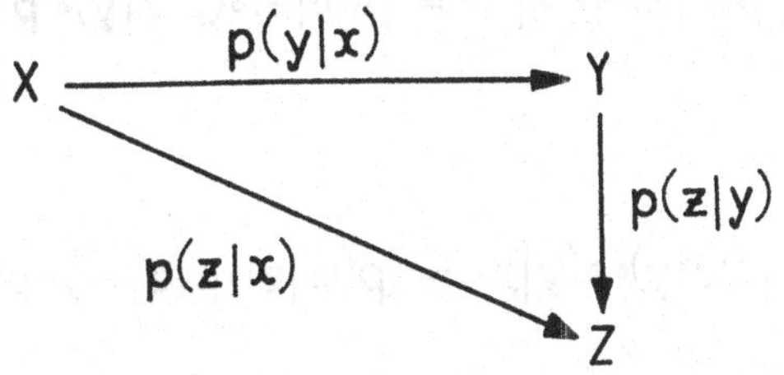

and in the form

$$X \xrightarrow{p(y|x)} Y \xrightarrow{p(z|y)} Z = X \xrightarrow{p(z|x)} Z\,.$$

In order to compose the random morphisms $p(y|x)$ and $p(z|y)$ the diagram given above must be completed by the random morphism $p(z|x)$.

PROPOSITION 1.2: <u>The composition law</u> $p(z|x) = p(z|y)\circ p(y|x)$ <u>defined by the equality (1.3) is associative.</u>

PROOF: Let us consider the objects X, Y, Z, W and three random morphisms

$$p(y|x) \in \mathrm{Hom}(X,Y)\,, \quad p(z|y) \in \mathrm{Hom}(Y,Z)\,, \quad p(w|z) \in \mathrm{Hom}(Z,W)\,.$$

From the following diagrams

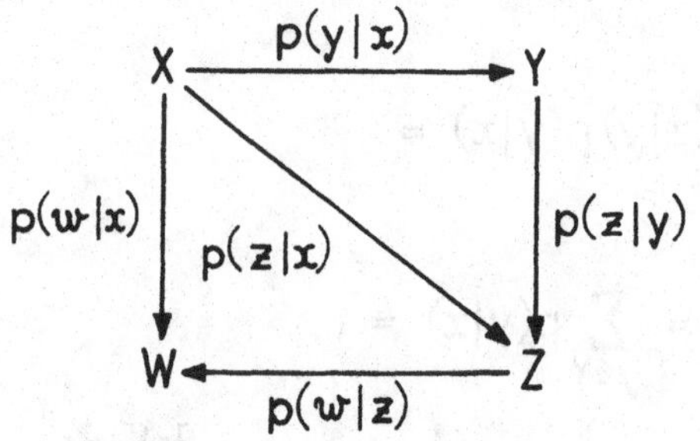

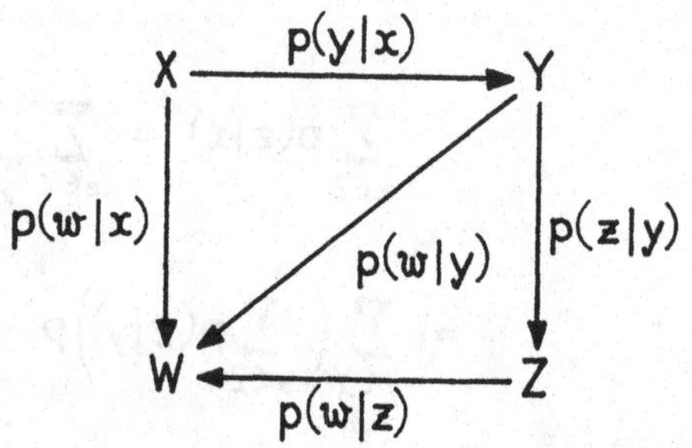

it follows that

$$p(w|x) = p(w|z)\circ p(z|x) = p(w|z)\circ(p(z|y)\circ p(y|x))$$

and respectively

$$p(w|z) = p(w|y)\circ p(y|x) = (p(w|z)\circ p(z|y))\circ p(y|x)\,.$$

Thus

$$p(w|z)\circ(p(z|y)\circ p(y|x)) = (p(w|z)\circ p(z|y))\circ p(y|x)$$

because their common value is

$$p(w|x) = \sum_{y\in Y}\sum_{z\in Z} p(w|z)p(z|y)p(y|x). \qquad \text{q.e.d.}$$

PROPOSITION 1.3: <u>For any two random morphisms</u> $p(y|x)\in \mathrm{Hom}(X,Y)$ $\tilde{p}(x|y)\in \mathrm{Hom}(Y,X)$ <u>we have the equality</u>

$$X \xrightarrow{1_X} X \xrightarrow{p(y|x)} Y = X \xrightarrow{p(y|x)} Y$$

and respectively

$$Y \xrightarrow{\tilde{p}(x|y)} X \xrightarrow{1_X} X = Y \xrightarrow{\tilde{p}(x|y)} X .$$

PROOF: According to the relations (1.3) and (1.5) if $p(x'|x)$ is the identical morphism we have

$$p(y|x')\circ p(x'|x) = \sum_{x'\in X} p(y|x')p(x'|x) = \sum_{x'\in X} p(y|x')\delta_{x',x} = p(y|x)$$

for every $x\in X$ and $y\in Y$. Also if $p(x|x')$ is the identical morphism we have

$$p(x|x')\circ\tilde{p}(x'|y) = \sum_{x'\in X} p(x|x')\tilde{p}(x'|y) = \sum_{x'\in X} \delta_{x',x}\tilde{p}(x'|y) = \tilde{p}(x|y) \qquad \text{q.e.d.}$$

Therefore, an FR-category of $\sum$-type is an associative category with identical morphisms whose objects are finite sets and whose morphisms are stochastic matrices, the law of composition of the morphisms and the identical morphisms being given by the equalities (1.3) and (1.5) respectively. Obviously an arbitrary morphism is not necessarily a usual function in an FR category of $\sum$ type but rather a random correspondence between two sets. According to this random correspondence, to every element $x \in X$ corresponds the element $y \in Y$ with the probability $p(y|x)$. The number $p(y|x)$ represents the transition probability from x to y. By means of the random morphism $p(y|x)$ to an arbitrary element $x \in X$ corresponds the arbitrary element y belonging to the set Y, with the probability $p(y|x)$. Of course, one or more elements $y' \in Y$ can exist so that $p(y'|x) = 0$. It is possible also to exist only one element from the set Y, denoted by y_x such that $p(y_x|x) = 1$ and obviously $p(y|x) = 0$ for every $y \in Y$, $y \neq y_x$.

Let $\mathcal{C}$ be an FR-category of $\sum$-type and $\mathcal{C}^0$ be the dual category. According to the definition of dual categories we have

$$\mathrm{Hom}_{\mathcal{C}^0}(X,Y) = \mathrm{Hom}_{\mathcal{C}}(Y,X)$$

for all objects X,Y in $\mathrm{Ob}(\mathcal{C}) = \mathrm{Ob}(\mathcal{C}^0)$ i.e.

$$p_{\mathcal{C}^0}(y|x) = p_{\mathcal{C}}(x|y).$$

Thus we have

$$0 \leqslant p_{\mathcal{C}^0}(y|x) \leqslant 1$$

for every $x \in X$, $y \in Y$ and

$$\sum_{x\in X} p_{\mathcal{C}^0}(y|x) = \sum_{x\in X} p_{\mathcal{C}}(x|y) = 1 .$$

Therefore $p_{\mathcal{C}^0}(y|x)$ is not necessarily a stochastic matrix and generally the dual of one FR-category of $\sum$-type is not also an FR-category of $\sum$-type. It is evident that if we have an FR-category of $\sum$-type whose morphisms are given by double-stochastic matrices, then every morphism

$$X \xrightarrow{p(y|x)} Y$$

satisfies the conditions

$$p(y|x) \geqslant 0 \text{ for every } x\in X ,\ y\in Y$$
$$\sum_{y\in Y} p(y|x) = \sum_{x\in X} p(y|x) = 1 \tag{1.6}$$

the dual category $\mathcal{C}^0$ is also an FR-category of $\sum$-type.

It is therefore possible to have morphisms defined by usual functions, in an FR-category of $\sum$ type. That is $u\in \mathrm{Hom}(X,Y)$ where $u: X \longrightarrow Y$ is a usual function. This morphism is a particular case of the random correspondence, namely the correspondence

$$X \xrightarrow{p_u(y|x)} Y$$

where

$$(1.7) \qquad p_u(y|x) = \begin{cases} 1, & \text{if} \quad y = u(x) \\ 0, & \text{if} \quad y \neq u(x) . \end{cases}$$

Let $\mathcal{C}$ be an FR-category of $\sum$-type containing a special object, namely a set having a single element, i.e.

$$\{e\} \in Ob(\mathcal{C}) .$$

A random morphism

$$p(x|e) \in Hom(\{e\}, X)$$

which may be denoted, without any possibility of confusion, by $p(x)$ induces a random distribution on the set X . $\{e\}$ is the <u>standard object</u>, $p(x)$ is the <u>standard morphism</u> and $\mathcal{C}$ is the <u>FR-category of $\sum$-type with standard object.</u>

If we consider also a random morphism $p(y|x) \in Hom(X, Y)$ then this morphism together with the standard morphism $p(x)$ defined above induce a random distribution $p(y)$ on the set Y , obtained by closing the following diagram

$$\begin{array}{ccc} \{e\} & \xrightarrow{p(x)} & X \\ & {\scriptstyle p(y)} \searrow & \downarrow {\scriptstyle p(y|x)} \\ & & Y \end{array}$$

where

$$p(y) = p(y|e) = \sum_{x\in X} p(y|x)\,p(x|e) = \sum_{x\in X} p(y|x)\,p(x) \;. \qquad (1.8)$$

Let us now define projection morphism. To construct it let us consider two objects X, Y belonging to an FR-category $\mathcal{C}$ of $\sum$-type. We shall define projection morphism (or projector) on the set X and we shall denote it by pr_X , the random morphism

$$pr_X = p(x'|x,y) \in \mathrm{Hom}(X \times Y\,, X) \qquad (1.9)$$

which is defined by the equalities

$$p(x'|x,y) = \begin{cases} 1 & \text{if} \quad x' = x \\ 0 & \text{if} \quad x' \neq x \end{cases}$$

for every $y \in Y$.

Analogously, we shall define the projection morphism on Y , and we shall denote it by pr_Y, the random morphism

$$pr_Y = p(y'|x,y) \in \mathrm{Hom}(X \times Y, Y) \qquad (1.10)$$

which is defined by the equality

$$p(y'|x,y) = \begin{cases} 1 & \text{if} \quad y' = y \\ 0 & \text{if} \quad y' \neq y \end{cases}$$

for every $x \in X$.

If we have the random morphism

$$Z \xrightarrow{\;p(x,y|z)\;} X \times Y$$

and the projection morphism on X , pr_X , then by closing the diagram

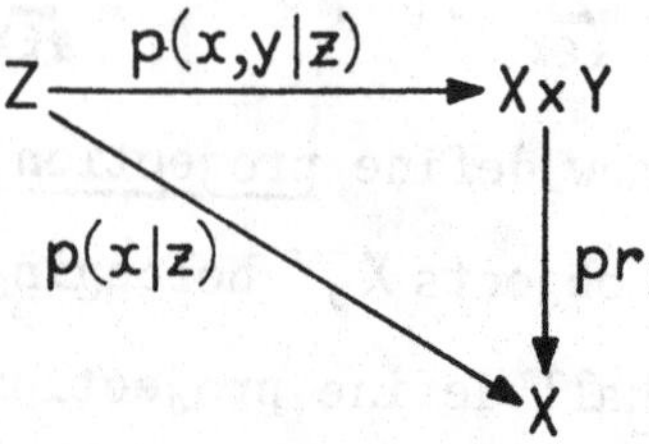

we shall obtain, according to (1.9), the random morphism

$$p(x|z) = \sum_{(x',y)\in X\times Y} p(x|x',y)\,p(x',y|z) = \sum_{y\in Y} p(x,y|z) .$$

Similarly, composing the given random morphism $p(x,y|z)$ with the projection morphism on Y , pr_Y , the random morphism

$$p(y|z) = pr_Y \circ p(x,y|z) = \sum_{x\in X} p(x,y|z)$$

will be obtained.

Particularly, if the standard morphism $p(x,y)\in \mathrm{Hom}(\{e\},X\times Y)$ is given, then, composing this morphism with the projection morphism pr_X, the standard morphism $p(x)$, is obtained where

$$p(x) = \sum_{y\in Y} p(x,y) .$$

Similarly, we have

$$pr_Y \circ p(x,y) = p(y)$$

where

$$p(y) = \sum_{x \in X} p(x,y) .$$

More explicitly, if we compose the standard morphism $p(x,y)$ with the projection morphisms pr_X and pr_Y respectively, we obtain just the standard morphisms $p(x)$ and $p(y)$ respectively, i.e. the marginal distributions of the bidimensional random distribution $p(x,y)$.

We shall call FR-category $\mathcal{C}$ of $\sum$-type with projection morphisms an FR-category of $\sum$-type so that for every two objects of the category $X,Y \in Ob(\mathcal{C})$ we have

$$pr_X \in Hom_{\mathcal{C}}(X \times Y, X) \ , \quad pr_Y \in Hom_{\mathcal{C}}(X \times Y, Y) .$$

Let us consider now an FR-category $\mathcal{C}$ of $\sum$-type with standard object $\{e\}$ and let $p_i(y|x) \in Hom(X,Y), (i=1,2,\dots,n)$ be n random morphisms of this category. We shall call the product morphism of the morphisms $p_i(y|x)$, $(i = 1,2,\dots,n)$ given above, the morphism

$$p(y_1,\dots,y_n|x_1,\dots,x_n) \in Hom(X \underbrace{\times \dots \times}_{n} X , Y \underbrace{\times \dots \times}_{n} Y)$$

being defined by the equality

$$(1.11)\qquad p(y_1,\dots,y_n|x_1,\dots,x_n) = p_1(y_1|x_1)\dots p_n(y_n|x_n)$$

for every $x_i \in X$, $y_i \in Y$, $(i = 1,2,\dots,n)$.

If, in particular, $p_i(y|x) = p(y|x)$ for every $i = 1,2,\dots,n$ and every $x\in X, y \in Y$ the product morphism $p(y_1,\dots,y_n|x_1,\dots,x_n)$ given above by the equality (1.11) is called the <u>product morphism of order n generated by the random morphism</u> $p(y|x)$

Let there be the random morphisms $p_i(x'|x) \in \mathrm{Hom}(X,X)$, $(i = 1,\dots,n-1)$ and the standard morphism $p(x)$. We shall call the standard morphism

$$p(x_0,x_1,\dots x_{n-1}) \in \mathrm{Hom}(\{e\}, \underbrace{X x \dots x X}_{n})$$

defined by the equality

$$p(x_0,x_1,\dots,x_{n-1}) = p(x_0)p_1(x_1|x_0)p_2(x_2|x_1)\dots p_{n-1}(x_{n-1}|x_{n-2})$$
(1.12)

for every $x_i \in X$, $(i = 0,1,\dots,n-1)$ as a <u>Markov morphism</u> generated by the random morphism given above.

Therefore, let $p(x'|x)$ be a random morphism preserving the standard morphism $p(x)$ with respect to the composition law of random morphisms, i.e.

$$\sum_{x\in X} p(x'|x)p(x) = p(x')$$

where in the right side of the last equality $p(x')$ is simply the given random morphism $p(x)$.

If $p_i(x'|x) = p(x'|x)$ for every $i = 1,\dots,n-1$ and

every $x \in X, x' \in X$ then the random morphism defined by the equality (1.12) is called the Markov morphism of order $(n-1)$ generated by the morphisms $p(x)$ and $p(x'|x)$.

It is necessary to notice here that throughout the paper, when referring to a Markov morphism of some order generated by some morphisms $p(x)$ and $p(x'|x)$ we shall suppose that the composition law applied to $p(x'|x)$ and $p(x)$ gives us just the standard morphism $p(x)$, i.e. $p(x'|x) \circ p(x) = p(x')$ where $p(x')$ is simply the morphism $p(x)$.

Let $\mathcal{C}$ be an FR-category of Σ-type with standard object and with projection morphisms. Let there be the random morphisms $p(x,y) \in Hom(\{e\}, X \times Y)$ and $p(y) \in Hom(\{e\}, X)$ such that $p(y) > 0$ for every $y \in Y$ where

$$p(y) = \sum_{x \in X} p(x,y) .$$

Let the reciprocal morphism corresponding to $p(x,y)$ and $p(y)$ be called the random morphism $p(x|y) \in Hom(Y, X)$ defined by the equality

$$p(x|y) = \frac{p(x,y)}{p(y)} . \tag{1.13}$$

Analogously, it is possible to define the reciprocal morphism $p(y|x)$ corresponding to $p(x,y)$ and $p(x)$, when $p(x) > 0$ for every $x \in X$ where

$$p(x) = \sum_{y \in Y} p(x,y) .$$

Obviously, it is possible to define a reciprocal morphism corresponding to the random morphisms $p(x,y)$ and $p(y), p(y) = \sum_{x\in X} p(x,y)$ even if $p(y)$ is not strictly positive, to be the random morphism $p(x|y) \in Hom(Y^*, X)$ defined by the equality (1.13) for every $x \in X$ and $y \in Y^*$ where $Y^* = \{y | y \in Y, p(y) \neq 0\}$. Generally we shall speak about the reciprocal morphism $p(x|y)$ (or $p(y|x)$) corresponding to the random morphisms $p(x,y)$ and $p(y)$, where $p(y) = \sum_{x\in X} p(x,y)$ (or to the random morphisms $p(x,y)$ and $p(x)$, where $p(x) = \sum_{y\in Y} p(x,y)$) only if $p(y) > 0$ for every $y \in Y$ (or $p(x) > 0$ for every $x \in X$).

If the following random morphisms are given, namely

$$\{e\} \xrightarrow{p(x)} X \,, \quad X \xrightarrow{p(y|x)} Y$$

we shall refer to the <u>Bayes morphism generated by</u> $p(y|x)$ <u>and</u> $p(x)$ to be as the standard morphism

$$\{e\} \xrightarrow{p(x,y)} X \times Y$$

defined by the equality

$$p(x,y) = p(y|x)\, p(x) \,. \tag{1.13'}$$

Then it is easy to construct the standard morphism

$$p(y) = \sum_{x\in X} p(x,y)$$

and to consider the reciprocal morphism $p(x|y)$ corresponding to $p(x,y)$ and $p(y)$, i.e.

$$p(x|y) = \frac{p(x,y)}{p(y)} .$$

From this last equality it is obvious that $p(x,y)$ is also the Bayes morphism generated by $p(x|y)$ and $p(y)$. According to (1.13) and (1.13') we can write

$$p(x|y) = \frac{p(y|x)\,p(x)}{\sum\limits_{x\in X} p(y|x)\,p(x)}$$

and we shall say that $p(x|y)$ is both the reciprocal morphism corresponding to the standard morphisms $p(x,y), p(y)$ and the <u>reciprocal morphism corresponding to the random morphisms</u> $p(y|x)$, $p(x)$. Analogously, we shall regard $p(y|x)$ as being both reciprocal morphism corresponding to the pair of standard morphisms $p(x,y)$, $p(x)$ and the reciprocal morphism corresponding to the pair of random morphisms $p(x|y)$, $p(y)$.

Of course, if the following random morphisms are given

$$p(x_1) : \{e\} \longrightarrow X_1 ; \quad p(x_2|x_1) : X_1 \longrightarrow X_2 ;$$

$$p(x_3|x_1,x_2) : X_1 \times X_2 \longrightarrow X_3$$

$$p(x_4|x_1,x_2,x_3) : X_1 \times X_2 \times X_3 \longrightarrow X_4$$

. . . .

$$p(x_n|x_1,x_2,\ldots,x_{n-1}) : X_1 \times X_2 \times \ldots \times X_{n-1} \longrightarrow X_n$$

then it is easy to construct the successive Bayes morphisms

$$p(x_1,x_2) = p(x_2|x_1)p(x_1):\{e\} \longrightarrow X_1 \times X_2$$

$$p(x_1,x_2,x_3) = p(x_3|x_1,x_2)p(x_1,x_2):\{e\} \longrightarrow X_1 \times X_2 \times X_3$$

$$p(x_1,x_2,x_3,x_4) = p(x_4|x_1,x_2,x_3)p(x_1,x_2,x_3):\{e\} \longrightarrow X_1 \times X_2 \times X_3 \times X_4$$

. . . .

$$p(x_1,x_2,\ldots,x_n) = p(x_n|x_1,x_2,\ldots,x_{n-1})p(x_1,x_2,\ldots,x_{n-1}):\{e\} \longrightarrow X_1 \times \ldots \times X_n.$$

Obviously

$$p(x_1,x_2,\ldots,x_n) = p(x_1)p(x_2|x_1)p(x_3|x_1,x_2)\ldots p(x_n|x_1,x_2,\ldots,x_{n-1})$$

shall be termed the Bayes morphism generated by the successive random morphisms $p(x_1), p(x_2|x_1), p(x_3|x_1,x_2),\ldots,p(x_n|x_1,\ldots,x_{n-1})$.

Let us consider the random morphism $p(y|x) \in Hom(X,Y)$ and an arbitrary partition $\pi(Y)$ of the set Y. Obviously, $\pi(Y) \subset \mathcal{P}(Y)$ where $\mathcal{P}(Y)$ is the set of all parts (subsets) of the set Y. Let us define the random morphism

$$p(\tilde{y}|x) \in Hom(X,\pi(Y))$$

defined by the equality

$$p(\tilde{y}|x) = \sum_{y \in \tilde{y}} p(y|x)$$

for every $x \in X$ and $\tilde{y} \in \pi(Y)$. The morphism $p(\tilde{y}|x)$ defined above will be called the extension of the morphism $p(y|x)$ to the partition $\pi(Y)$ or the canonical morphism of transition to the partition $\pi(Y)$.

Let us consider again the morphism $p(y|x) \in Hom(X,Y)$. Then, obviously, all the restrictions of this morphism to the subsets $X' \subset X$ i.e. the random morphisms $p(y|x') \in Hom(X',Y)$ for all $X' \subset X$ where $p(y|x')$ coincides with the morphism $p(y|x)$ for every $x \in X'$ are very well-defined. For every $x_0 \in X$ in particular, the restriction of the morphism $p(y|x)$ to the set $\{x_0\}$ i.e. the random morphism $p(y|x_0) \in Hom(\{x_0\}, Y)$ is also well-defined.

The reciprocal morphisms, the Markov morphisms, the Bayes morphisms, the product morphisms, the canonical morphisms of transition to partitions and the restrictions of the morphisms to the various subsets of the sources will be classed together as derived morphisms. An FR-category of $\sum$-type having standard object, projection morphisms and containing derived morphisms will be called FR-category of $\sum$-type well-equipped.

Let $\mathcal{C}$ be an FR-category of $\sum$-type well-equipped. The subsets of $\mathcal{M}(\mathcal{C})$ containing those random morphisms from which by composition or derivation (the construction of the reciprocal morphisms, Markov morphisms, Bayes morphisms or product morphisms) it is possible to obtain all the random morphisms of the category $\mathcal{C}$ will be termed the set of elementary morphisms of the above mentioned category. If we eliminate the projection morphisms and the identical morphisms from the set of elementary morphisms we shall obtain the so-called set of primary morphisms. Therefore, in defining an FR-category of $\sum$-type well-equipped, it will be sufficient to indicate both the primary morphisms and the objects related to these morphisms, i.e. the sources and the endings of the primary morphisms. The sources and the endings of the derived morphisms or of the morphisms obtained by composing the morphisms will belong to the set of the objects of the respective category. This last supposition is made only to simplify the writing.

Let $\mathcal{C}$ now be an FR-category of $\sum$-type having the standard object $\{e\}$. If we compose the random morphisms $p(x) \in \mathrm{Hom}(\{e\}, X)$ and $p(y|x) \in \mathrm{Hom}(X, Y)$ we obtain the standard morphism $p(y) \in \mathrm{Hom}(\{e\}, Y)$ defined by the equality (1.8). We shall call the morphism $p(x,y|x') \in \mathrm{Hom}(X, X \times Y)$ defined by the equality

$$(1.14) \quad p(x,y|x') = p(y|x')\delta_{xx'} = \begin{cases} p(y|x') & \text{if} \quad x = x' \\ 0 & \text{if} \quad x \neq x' \end{cases}$$

as <u>the extension of the morphism</u> $p(y|x)$ <u>to the product set.</u> This last morphism composed with the standard morphism $p(x) \in \mathrm{Hom}(\{e\}, X)$ gives us

$$p(x,y|x') \circ p(x') = \sum_{x' \in X} p(x,y|x')p(x') =$$

$$= \sum_{x' \in X} p(y|x')\delta_{x'x}p(x') = p(y|x)p(x) = p(x,y)$$

where $p(x,y)$ is the Bayes morphism generated by the morphisms $p(x)$ and $p(y|x)$.

PROPOSITION 1.5: _If_ $\mathcal{C}$ _is an FR-category of_ Σ_-type well-equipped and if_ $p(x)\in Hom_{\mathcal{C}}(\{e\},X)$, $p(x)>0$ _for every_ x _and_ $p(y|x)\in Hom_{\mathcal{C}}(X,Y)$ _then the extension of the morphism_ $p(y|x)$ _to the product set is a morphism of the category_ $\mathcal{C}$ _too, namely_

$$p(x,y|x')\in Hom_{\mathcal{C}}(X,X\times Y)$$

PROOF: Since the category $\mathcal{C}$ is well-equipped, if $p(x)\in\mathcal{M}(\mathcal{C})$ and $p(y|x)\in\mathcal{M}(\mathcal{C})$ then $p(x,y)=p(y|x)p(x)\in\mathcal{M}(\mathcal{C})$ and $p(x'|x,y)\in\mathcal{M}(\mathcal{C})$ where the last is the projection morphism $p(x'|$ $|x,y)\in Hom(X\times Y,X)$ defined by (1.9) such that

$$p(x') = p(x'|x,y)\circ p(x,y)$$

Let us consider the derived morphism

$$p(x',x,y) = p(x'|x,y)p(x,y) = \delta_{x',x}p(x,y) = \begin{cases} p(x,y) & \text{if} \quad x=x' \\ 0 & \text{if} \quad x\neq x' \end{cases}$$

Obviously $p(x',x,y)\in\mathcal{M}(\mathcal{C})$ or more precisely

$$p(x',x,y)\in Hom_{\mathcal{C}}(\{e\},X\times X\times Y)\ .$$

Because

$$p(x') = \sum_{x\in X}\sum_{y\in Y} p(x',x,y) = \sum_{y\in Y} p(x',y)$$

is strictly positive and $\mathcal{C}$ is well-equipped, the reciprocal morphism $p(x,y|x')$ corresponding to $p(x',x,y)$ and $p(x')$ is also a morphism of the category $\mathcal{C}$ but, from the equality (1.13) we have

$$p(x,y|x') = \frac{p(x',x,y)}{p(x')} = \frac{\delta_{xx'}p(x,y)}{p(x')} = \begin{cases} p(y|x') & \text{if} \quad x = x' \\ 0 & \text{if} \quad x \neq x' \end{cases}$$

therefore we even obtain the extension of the morphism $p(y|x)$ to the product set. q.e.d.

When speaking about the extension of the morphism $p(y|x)$ to the product set it will always be presumed that the standard morphism $p(x)$ exists and that $p(x) > 0$ for every $x \in X$.

Let us examine the various possibilities for the evaluation of the random morphism $p(x,y,z|x') \in Hom(X, X\times Y\times Z)$ obtained by composing the extensions to the product set $p(x,y|x')$ and $p(x,y,z|x',y')$ respectively, of the random morphisms $p(y|x) \in Hom(X,Y)$ and $p(z|x,y) \in Hom(X\times Y, Z)$.

PROPOSITION 1.6 : <u>We have</u>

$$p(x,y,z|x') = p(z|x,y)p(y|x')\delta_{x'x}. \tag{1.15}$$

PROOF: According to the definition (1.14) of the extensions of the random morphisms to the product set we have

$$p(x,y,z|x') = p(x,y,z|x'',y'')\circ p(x'',y''|x') =$$

$$= \sum_{(x'',y'')\in X\times Y} p(x,y,z|x'',y'')p(x'',y''|x') = \sum_{x''\in X}\sum_{y''\in Y} p(z|x'',y'')\delta_{(x'',y''),(x,y)}p(y''|x')\delta_{x'',x'} =$$

$$= p(z|x,y)p(y|x')\delta_{x',x} .$$

q.e.d.

Obviously, from (1.14) it follows that the extension to the product set of the morphism $p(y,z|x) \in \mathrm{Hom}(X, Y \times Z)$ is the morphism

$$p(x,y,z|x') \in \mathrm{Hom}(X, X \times Y \times Z)$$

defined by the equality

$$p(x,y,z|x') = p(y,z|x')\delta_{x',x} . \tag{1.16}$$

If the morphisms

$$p(x_k|x_1,\ldots,x_{k-1}) \in \mathrm{Hom}(X^{k-1}, X) , \quad (k = 2,\ldots,n)$$

are given where $X^k = \underbrace{X \times \ldots \times X}_{k}$, and if the random morphism

$$p(x_1,\ldots,x_{k-1},x_k|x'_1,\ldots,x'_{k-1}) \in \mathrm{Hom}(X^{k-1}, X^k)$$

represents the extension to the product set of the morphism

$$p(x_k|x_1,\ldots,x_{k-1})$$

for every $k = 2,3,\ldots,n$, then, by mathematical induction from the last proposition we have

$$p(x_1,\ldots,x_n|x') = p(x_1,\ldots,x_n|x_1^{(n-1)},\ldots,x_{n-1}^{(n-1)}) . \tag{1.17}$$

$$\circ p(x_1^{(n-1)},\dots,x_{n-1}^{(n-1)}|x_1^{(n-2)},\dots,x_{n-2}^{(n-2)})\circ\dots$$

(1.17)

$$\dots\circ p(x_1^{(3)},x_2^{(3)},x_3^{(3)}|x_1^{(2)},x_2^{(2)})\circ p(x_1^{(2)},x_2^{(2)}|x') .$$

Composing this last morphism $p(x_1,\dots,x_n|x')\in Hom(X,X^n)$ with the standard morphism $p(x)\in Hom(\{e\},X)$ we shall obtain the morphism

(1.18)

$$p(x_1,\dots,x_n) = p(x_1,\dots,x_n|x')\circ p(x') =$$
$$= p(x_n|x_1,\dots,x_{n-1})p(x_{n-1}|x_1,\dots,x_{n-2})\dots p(x_3|x_1,x_2)p(x_2|x_1)p(x_1)$$

It can be said that the morphism $p(y|x)\in Hom(X,Y)$ is strictly deterministic if for every $x\in X$ there exists $y_x\in Y$ so that $p(y_x|x) = 1$. If $0<\varepsilon<1$ so that $p(y_x|x)>1-\varepsilon$, then the random morphism $p(y|x)\in Hom(X,Y)$ is termed as being ε-deterministic, or ε-injection.

Obviously, an ε-deterministic morphism is a random morphism that may be considered as a strictly deterministic morphism with an error smaller than ε .

It can be stated that an FR-category $\mathcal{C}$ of Σ-type is complete if for every sequence of random morphisms $p_n(y|x)\in Hom(X,Y)$, $(n=1,2,\dots)$ for which the limit $\lim_{n\to\infty} p_n(y|x)$ exists, whichever be $x\in X, y\in Y$ having either the form $p(y|x)$ or the form $p(y)$ we have $p(y|x)\in Hom(X,Y)$ or $p(y)\in Hom(\{e\},Y)$ for every two objects X, Y from $Ob(\mathcal{C})$.

Let us return to the name. In the name "FR-category of Σ-type" the adjective "finite" refers to the objects of the category which are finite sets. The adjective "random" refers to the morphisms of the category which are stochastic (random) matrices whose elements are probabilities, i.e. real numbers belonging to the interval $[0,1]$. The expression "of Σ-type" was added to underline the fact that the sum-procedure was utilized in the definition of the composition of the random morphisms.

It is possible to introduce several kinds of finite random categories, four of which will be defined as follows:

<u>FR-category of M-type</u> may be defined as one in which

a) The objects are finite sets;

b) The morphisms are matrices with elements belonging to the unit interval $[0,1]$ of the real line (i.e. they may be considered as probabilities). For every pair of objects X, Y if $Hom(X,Y) \neq \emptyset$ then, an arbitrary random morphism $u \in Hom(X,Y)$ has the form

$$u = (p(y|x))_{\substack{x \in X \\ y \in Y}}$$

or simply $u = p(y|x)$ where $0 \leq p(y|x) \leq 1$ for every $x \in X, y \in Y$ having the following interpretation: the arbitrary element $y \in Y$ corresponds to the arbitrary element $x \in X$ with the probability given by the number $p(y|x)$.

c) For every system of three objects X, Y, Z the composition of the arbitrary random morphisms $p(y|x) \in Hom(X,Y)$ and $p(z|y) \in Hom(Y,Z)$ gives us the random morphism $p(z|x) \in Hom(X,Z)$ defined by the equality

$$p(z|x) = \underset{y \in Y}{Max}\, p(z|y)p(y|x). \tag{1.19}$$

Similarly an FR-category is of m-type, of Mm-type or of mM-type if in the condition c) given above, the composition law of the morphisms (1.19) is replaced respectively by

$$p(z|x) = \min_{y \in Y} p(z|y)p(y|x) \tag{1.20}$$

by

$$p(z|x) = \underset{y \in Y}{Max} \{\min(p(z|y), p(y|x))\} \tag{1.21}$$

or by

$$p(z|x) = \min_{y \in Y} \{Max(p(z|y), p(y|x))\}. \tag{1.22}$$

In these cases the name of the type of the category refers to the manner of composition of the random morphisms.

FR categories of $\sum$-type will be used almost constantly in the following chapters.

The term "FR-category" shall be taken to mean an FR-category of $\sum$-type, whilst "FR well-equipped" shall be understood as being FR-category of $\sum$-type well-equipped".

The finite random categories of $\sum^{*}$-type as defines by my student Maria Cristea, may be amplified as follows. A finite random category $\mathcal{C}$ is of $\sum^{*}$-type if:

a) its objects are finite sets;

b) The morphisms are families of random (stochastic) matrices.

Given $X, Y \in Ob(\mathcal{C})$ and $a = \text{card}X$ the following set can be introduced:

$$\Delta_a = \left\{(\alpha_1, \dots, \alpha_a) \mid 0 \leq \alpha_i \leq 1,\ \sum_{i=1}^{a} \alpha_i = 1\right\}.$$

If $Hom(X,Y) \neq \emptyset$ then an element of this set is a whole family of stochastic matrices for which the indices belong to the set Δ_a . Thus $\bar{\bar{p}}(y|x) \in Hom(X,Y)$ has the form

$$\bar{\bar{p}}(y|x) = (p_{\alpha_1,\dots,\alpha_a}(y|x))_{(\alpha_1,\dots,\alpha_a)\in\Delta_a}\ .$$

c) For every $X, Y, Z \in Ob(\mathcal{C})$ we define the composition of the random morphisms $\bar{\bar{p}}(y|x) \in Hom(X,Y)$, $\bar{\bar{p}}(z|y) \in Hom(Y,Z)$ as being the random morphism $\bar{\bar{p}}(z|x) \in Hom(X,Z)$ denoted by

$$\bar{\bar{p}}(z|x) = \bar{\bar{p}}(z|y) \circ \bar{\bar{p}}(y|x)$$

and constructed in the following way: if $a = \text{card}X$ and $b = \text{card}Y$ let Δ_a and Δ_b be the corresponding sets of indices. The random morphism

$$\bar{\bar{p}}(z|x) = (p_{\alpha_1,\dots,\alpha_a}(z|x))_{(\alpha_1,\dots,\alpha_a)\in\Delta_a}$$

is defined by the equality

$$p_{\alpha_1,\dots,\alpha_a}(z|x) = \sum_{y\in Y} p_{\alpha_1,\dots,\alpha_a}(y|x)\,p_{\beta_{x_1},\dots,\beta_{x_b}}(z|y)$$

whichever be $x\in X$, $z\in Z$ where $(\alpha_1,\dots,\alpha_a)\in\Delta_a$, $(\beta_{x_1},\dots,\beta_{x_b})\in\Delta_b$ here $\beta_{x_1},\dots,\beta_{x_b}$ representing the line x in the stochastic matrix $p_{\alpha_1,\dots,\alpha_a}(y|x)$.

d) For every $X\in Ob(\mathcal{C})$ we define the morphism $1_X^{\ell}\in Hom(X,X)$ composed by a single stochastic matrix. Thus

$$p_{\alpha_1,\dots,\alpha_a}(x'|x) = p(x'|x) = \delta_{x',x}$$

whichever be $(\alpha_1,\dots,\alpha_a)\in\Delta_a$ where $a = card\,X$.

It is easy to verify that the composition law just defined is associative and that for two random morphisms

$$\bar{\bar{p}}(x|y) = (p_{\beta_1,\dots,\beta_b}(x|y))_{(\beta_1,\dots,\beta_b)\in\Delta_b}\in Hom(Y,X)\,,\quad (b = card\,Y)$$

$$\bar{\bar{p}}(y|x) = (p_{\alpha_1,\dots,\alpha_a}(y|x))_{(\alpha_1,\dots,\alpha_b)\in\Delta_a}\in Hom(X,Y)\,,\quad (a = card\,X)$$

the random morphism $1_X^{\ell}(x'|x)\circ\bar{\bar{p}}(x|y)\in Hom(X,Y)$ has the elements

$$(p_{\beta_1,\dots,\beta_b}(x'|y))_{(\beta_1,\dots,\beta_b)\in\Delta_b}$$

i.e. just the morphism $\bar{\bar{p}}(x|y)$ and the random morphism

$$\bar{\bar{p}}(y|x)\circ 1_X^{\ell}(x|x')\in Hom(X,Y)$$

is the family consisting of a single stochastic matrix, namely

$$p_{\alpha_{x'_1},\dots,\alpha_{x'_a}}(y|x')$$

where $\alpha_{x'_1},\dots,\alpha_{x'_a}$ is the line x' in the stochastic matrix $1_X^{\ell}(x|x')$. Therefore $1_X^{\ell}(x'|x)$ is <u>left</u> identical morphism of the given category.

Of course, for an arbitrary random morphism

$$\bar{\bar{p}}(y|x) = (p_{\alpha_1,\dots,\alpha_a}(y|x))_{(\alpha_1,\dots,\alpha_a)\in\Delta_a} \in \mathrm{Hom}(X,Y)$$

($a = \mathrm{card}\,X$) any element of the set Δ_a is a random distribution on the set X, therefore in the FR-categories of $\sum^*$-type for every possible random distribution on the source of the random morphism we can pick up a corresponding transition stochastic matrix representing this random morphism. The random morphisms of this type correspond to the stochastic evolution with complete connexions (in the sense of Onicescu and Mihoc). We obtain a Markov classical stochastic evolution if every family defining a random morphism is composed of a single transition matrix, i.e.

$$p_{\alpha_1,\dots,\alpha_a}(y|x) = p(y|x)$$

for every $(\alpha_1,\dots,\alpha_a)\in\Delta_a$, ($a = \mathrm{card}\,X$). In this case we again obtain the FR-categories of $\sum$-type.

Chapter 2

EXAMPLES OF FINITE RANDOM CATEGORIES

Let us give some examples of FR-categories. Almost all will be FR-categories of Σ-type, well-equipped and therefore, according to the theory developed in the first chapter, their definition requirres only the primary morphisms and the corresponding objects, i.e. the sources and the endings of the primary morphisms. The sets which are sources or endings both of the derived morphisms and of the morphism obtained by composition of the morphisms will belong implicitly to the class of objects of respective category.

For an FR-category of Σ-type which is not well-equipped we shall suppose that the identical morphisms corresponding to the objects of the category are given too. All these suppositions are made to simplify the notations. Let us now consider the desired examples.

1. Discrete-memory-less noisy communication channel.

An arbitrary communication channel is characterized by the set of signals at the entrance of the channel (input signals), the set of signals at the exit of the channel (output signals) and the amount of perturbation (noise) altering the transmission of the signals through the channel. The noise may be characterized only from the statistical point of view. As a matter of fact, each possible output signal y with the probability $p(y|x)$ may correspond to an arbitrary input signal x . Every $x, p(y|x)$ as function of the output signal y , is a probability measure on the set of output signals of the respective communication channel characterizing (from the statistical point of view) the perturbation of the given channel.

A noisy-discrete-memory-less communication channel is an FR-category of $\sum$-type, i.e.

$$\mathrm{Ob}(\mathscr{C}) = \{X, Y\}$$

$$\mathrm{Hom}(X, Y) = \{p_n(y|x), (n = 1, 2, \ldots)\}$$

where X represents the set of input signals; Y represents the set of output signals and $p_n(y|x)$ is the perturbation at the moment n, i.e. the probability of receiving (as a consequence of the presence of the noise on the communication channel) the output signal y at the end of the channel, at the moment n if the input signal x were transmitted through the channel, whichever be $x \in X$ and $y \in Y$. If $p_n(y|x) = p(y|x)$ the communication channel is stationary.

2. Discrete memory-less-noisy communication system.

A communication system is obtained by coupling an information source with a communication channel. A discrete-memory-less noisy communication system is an FR-category of $\sum$-type well-equipped such that

$$\mathrm{Hom}(\{e\},X) = \{p_n(x), (n = 1,2,\ldots)\}$$

$$\mathrm{Hom}(X,Y) = \{p_n(y|x), (n = 1,2,\ldots)\}$$

where X represents the set of signals which are transmitted, Y the set of signals which are received (at the end of the communication channel), $p_n(x)$ is the probability of the input signal at the moment n, $p_n(y|x)$ is the probability of receiving the output signal y at the moment n, if at the same moment the output signal x were transmitted. If $p_n(x) = p(x)$ and $p_n(y|x) = p(y|x)$ the communication system is stationary.

3. Finite abstract random automaton.

A finite abstract random automaton is an FR-category of $\sum$-type well-equipped and complete, i.e.

$$\mathrm{Hom}(\{e\},A\times X) = \{p_0(a,x)\}$$

$$\mathrm{Hom}(A\times X,A\times X) = \{p_n(a',x'|a,x), (n = 1,2,\ldots)\} \tag{2.1}$$

$$\mathrm{Hom}(A\times X,Y) = \{p_n(y|a,x), (n = 1,2,\ldots)\}$$

where: A represents the set of states of the automaton, and X the set of input signals. Y represents the set of output signals and $p_0(a,x)$ initial probability of the state a and of the input signal x. $p_n(a',x'|a,x)$ represents the transition probability of the automaton from its state a with the input signal x at the moment $(n-1)$ to the state a' with the input signal x' at the next moment n ; $p_n(y|a,x)$ represents the probability of the output signal y at the moment n ,if at the previous moment the automaton was in the state a with the input signal x . Here $p_n(a',x'|a,x)$ is called the transition morphism at the moment n whereas $p_n(y|a,x)$ is called the output morphism at the same moment n .

If $p_n(a',x'|a,x) = p(a',x'|a,x)$ and $p_n(y|a,x) = p(y|a,x)$ for every n the random automaton is stationary. A stationary random automaton for which the composition law of the morphism $p(a',x'|a,x)$ with the morphism $p_0(a,x)$ which preserves the standard morphism $p_0(a,x)$ i.e.

$$p(a',x'|a,x) \circ p_0(a,x) = p_0(a',x')$$

is called an homogeneous random automaton.

4. Finite abstract fuzzy automaton

The fuzzy automata are such automata for which one, two or all the corresponding sets are fuzzy sets. At the same time the composition law of the morphism is different.

An optimistic (respectively pessimistic) finite abstract fuzzy automaton is an FR-category of Mm -type (respectively of mM -type) such that the morphisms are those indicated in the relations (2.1), together with the morphisms obtained by composing these morphisms in accordance to the composition law (1.22) (respectively to (1.21)).

5. Two-person games.

The game is a mathematical model for conflicts. A two-person game is an FR-category of $\sum$-type well-equipped such that

(2.2)

$$\mathrm{Hom}(\{e\},X) = \{p_1(x)\}$$

$$\mathrm{Hom}((X\times Y)^{i-1},X) = \{p_i(x|x_1,y_1,\dots,x_{i-1},y_{i-1})\} \quad (i = 1,2,\dots,n)$$

$$\mathrm{Hom}((X\times Y)^{j-1}\times X,Y) = \{p_j(y|x_1,y_1,\dots,x_{j-1},y_{j-1},x_j)\}, \quad (j = 1,\dots,n)$$

$$\mathrm{Hom}((X\times Y)^n,U_k) = \{p_k(u|x_1,y_1,\dots,x_n,y_n) = \delta_{u,u_k(x_1,y_1,\dots,x_n,y_n)}\} \ (k=1,2)$$

where: X represents the set of all possible actions of the first player, Y the set of all possible actions of the second player and n represents the number of the actions executed by each player, i.e. the duration of the game; U_k is a finite set of non-negative real numbers representing the utilities of the different variants of the game for the player $k(k=1,2)$. $p_1(x)$ is the probability of the first action of the first player; $p_i(x|x_1, y_1,\dots,x_{i-1},y_{i-1})$ is the probability of the action x of the first player at the moment i of the game (a moment of the game is composed by one action of the first player followed by one action of the second player) if the successive actions at the previous moments were $x_1,\dots,x_{i-1}$ for the first player and $y_1,\dots,y_{i-1}$ for the second one for every $i=2,\dots,n$. The morphism $p_j(y|x_1,y_1,\dots,x_{j-1},y_{j-1},x_j)$ is the probability of the action y of the second player at the moment j of the game if at the previous moments the successive actions were $x_1,\dots,x_j$ for the first player and $y_1,\dots,y_{j-1}$ for the second one. $u_k(x_1,y_1,\dots, x_n,y_n)$ represents the utility for the player k $(k=1,2)$ of the variant of the game composed of the successive actions $x_1,y_1,\dots,x_n,y_n$ of the players.

Then

$$U_k=\{u_k(x_1,y_1,\dots,x_n,y_n)|x_i\in X,y_i\in Y,(i=1,\dots,n)\}\ ,\quad (k=1,2)$$

We have

$$p_k(u|x_1,y_1,\dots,x_n,y_n)=\begin{cases}1 & \text{if} \quad u=u_k(x_1,y_1,\dots,x_n,y_n)\\ 0 & \text{if} \quad u\neq u_k(x_1,y_1,\dots,x_n,y_n)\end{cases}$$

We shall describe every system $(x_1,x_2,\dots,x_n)$ composed of n actions of the first player, as a deterministic strategy of the first player. The same is true for the second player. A variant of the game is composed of one deterministic strategy of each player. A random strategy of one player is a system of probabilities with the sum equal to 1 defined on the set of all possible deterministic strategies of the respective player.

A two-person game will be called a two-person game with independent strategies if:

$$p_i(x|x_1,y_1,\dots,x_{i-1},y_{i-1}) = p_i(x|x_1,\dots,x_{i-1}), \quad (i = 2,3,\dots,n)$$

$$p_j(y|x_1,y_1,\dots,x_{j-1},y_{j-1},x_j) = p_j(y|y_1,\dots,y_{j-1}), \quad (j = 1,2,\dots,n).$$

6. Learning system

A learning system is an FR-category of Σ-type well-equipped and complete (the limit of every convergent sequence of random morphisms of the category belongs also to this category) such as

$$\mathrm{Hom}(\{e\},S) = \{p_1(s)\}, \quad \mathrm{Hom}(S,A) = \{p_1(a|s)\},$$

$$\mathrm{Hom}(S\times A,O) = \{p_n(o|s,a),(n = 1,2,\dots)\},$$

$$\mathrm{Hom}(S\times A\times O, S\times A) = \{p_{n+1}(s',a'|s,a,o),(n = 1,2,\dots)\}$$

in which S is the set of stimuli, A is the set of the organism's responses, O the set of results and $p_1(a|s)$ the probability of the response a in the first experiment conditioned by the stimulus s. $p_1(s)$ is the probability of the stimulus s during the first experiment whilst $p_n(o|s,a)$ is the probability of the result o in the experiment n, conditioned by the stimulus s and the response a from the same experiment $p_{n+1}(s',a'|s,a,o)$ is the probability of the stimulus s' and the response a' after the experiment n if in this experiment the stimulus s, the response a, and the result o all occurred together.

The learning system is stationary if the random morphisms are independent of n.

7. Predictive system

A predictive system is an FR-category of Σ-type well-equipped such that

$$\mathrm{Hom}(\{e\},\tilde{H}) = \{p_0(h)\}$$

$$\mathrm{Hom}(\tilde{H}\times D^{\ell-1},D) = \{p(d_{i_\ell}|h,d_{i_1},\dots,d_{i_{\ell-1}})\} \qquad (\ell = 1,\dots,k),$$

where k is a large natural number.

Here: $\tilde{H}$ is the set of available hypotheses and D the set of all possible outcomes (results). Also $p_0(h)$ represents the initial (or a priori) probabilities of the available hypotheses and themorphism $p(d_{i_\ell}|h,d_{i_1},d_{i_2},\dots,d_{i_{\ell-1}})$ represents the probability of the result d_{i_ℓ} at the moment ℓ conditioned both by the hypotheses h and by the successive previous results $d_{i_1}, d_{i_2}, \dots, d_{i_{\ell-1}}$ whichever be $\ell = 1,\dots,k$.

Chapter 3

PROCESSES IN FINITE RANDOM CATEGORIES

Let $\tilde{N}$ be a finite denumerable totally ordered set with prime element (this is often the set of natural numbers) and let $\mathcal{C}$ be an FR-category well-equipped. We shall call a process (with discrete time) in the category $\mathcal{C}$ any application.

$$P : \tilde{N} \longrightarrow \mathcal{P}(\mathcal{M}(\mathcal{C})) \tag{3.1}$$

(where $\mathcal{P}(A)$ represents the set of all parts of A)such that if i', i'' are two arbitrary successive elements of the set $\tilde{N}$ (i.e. $i' < i''$ and i does not exist so that $i' < i < i''$ where $<$ represents the order relation in $\tilde{N}$) then there exists at least one morphism $u \in P(i'')$ and at least one morphism $v \in P(i')$ in such a manner that the source of u coincides with the ending of v, i.e. $S(u) = E(v)$. The process P is finite or infinite if the set $\tilde{N}$ is finite or infinite.

A sequence ('sequence' here signifies a finite or denumerable set) of morphisms $(u_i)_{i \in \tilde{N}}, u_i \in P(i)$ so that for every two successive elements i', i'' from $\tilde{N}$ the source of $u_{i''}$ coinciding with the ending of $u_{i'}$ is called the skeleton of the process P.

PROPOSITION 3.1: Let A be an object of the category $\mathcal{C}$ and let P be a process (with discrete time) in the category $\mathcal{C}$,

$$P : \tilde{N} \longrightarrow \mathcal{P}(\mathcal{M}(\mathcal{C}))$$

Let $(u_i)_{i \in \tilde{N}}$ also be a skeleton of the process P. If for every $i \in \tilde{N}$ one morphism v_i^1 exists so that $S(v_i^1) = S(u_i)$, $E(v_i^1) = A$ together with one morphism v_i^2 so that $S(v_i^2) = A$, $E(v_i^2) = S(u_i)$ then the skeleton $(u_i)_{i \in \tilde{N}}$ together with the standard morphism $p(a) \in \mathrm{Hom}(\{e\}, A)$ induce a Markov chain on A.

PROOF: From the definition given above it follows that the compound morphism $v_{i''}^1 \circ u_{i'} \circ v_{i'}^2$ where i' and i'' are two arbitrary successive elements belonging to $\tilde{N}$, defines a passage probability from the moment i' to the moment i'' on the set A. When the successive elements i' and i'' take on all the values of the set $\tilde{N}$, these transition probabilities define a Markov chain on the set A with the initial distribution given by the standard morphism $p(a) \in \mathrm{Hom}(\{e\}, A)$. q.e.d.

If the skeleton $(u_i)_{i \in \tilde{N}}$ has the property that

$$u_i = u \ , \quad v_i^1 = v^1 \ , \quad v_i^2 = v^2 \tag{3.2}$$

for every $i \in \tilde{N}$ then the corresponding Markov chain induced on set A is stationary.

We shall say that process P defined by (3.1) is a process with the skeleton tied to the object A given the existence of a skeleton $(u_i)_{i\in\tilde{N}}$ of the process P and the morphisms v_i^1 and v_i^2 for every $i\in\tilde{N}$ so that

$$S(v_i^1) = S(u_i)\,, \quad E(v_i^1) = A\,, \quad S(v_i^2) = A\,, \quad E(v_i^2) = S(u_i)$$

Obviously, if i' and i'' are two successive elements of the set $\tilde{N}$ (i.e. two successive moments), the morphism $v_{i''}^1 \circ u_{i'} \circ v_{i'}^2$ has its source A and A as its ending. Therefore, by closing the diagram formed with arrows $v_{i''}^1, u_{i'}, v_{i''}^2$ we obtain

$$A \xrightarrow{p_{i',i''}(a'|a)} A\,.$$

Let us suppose that set $\tilde{N}$ is denumerable and let us denote explicitly

$$\tilde{N} = \{i_0, i_1, \ldots, i_n, i_{n+1}, \ldots\}\,.$$

Let us introduce the morphism

$$\begin{aligned} p_{i_0,i_n}(a'|a) = p_{i_{n-1},i_n}(a'|a_{n-1})\circ p_{i_{n-2},i_{n-1}}(a_{n-1}|a_{n-2})\circ\ldots \\ \ldots\circ p_{i_1,i_2}(a_2|a_1)\circ p_{i_0,i_1}(a_1|a)\,. \end{aligned} \tag{3.3}$$

Here $p_{i_0,i_n}(a'|a)$ represents the transition morphism corresponding to the interval $[i_0,i_n]$. The infinite process P with the skeleton tied to the object A is <u>mixing</u> with respect to A if there exists

$$\lim_{n\to\infty} p_{i_0,i_n}(a'|a) = p_\infty(a') \tag{3.4}$$

this limit being independent both of a and of i_0. Let us consider also a standard morphism $p_{i_0}(a) \in \mathrm{Hom}(\{e\}, A)$. By closing the diagram

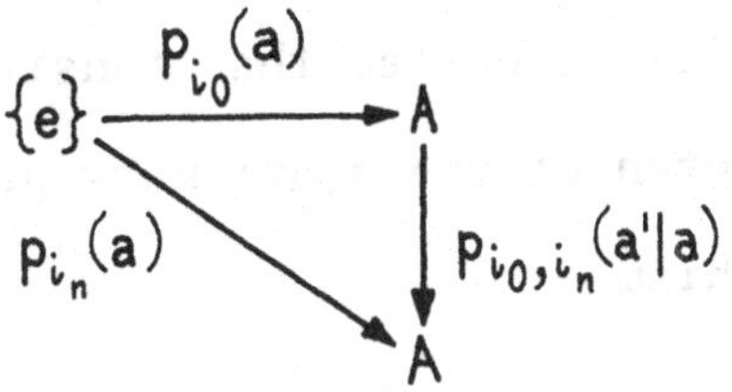

we obtain the standard morphism

$$p_{i_n}(a) = p_{i_0,i_n}(a'|a) \circ p_{i_0}(a) \tag{3.5}$$

and to this last morphism we attach the entropy

$$H_{i_n}(A) = -\sum_{a \in A} p_{i_n}(a) \log p_{i_n}(a) \tag{3.6}$$

as the measure of uncertainty contained in the object A at the moment i_n of the process.

We shall say that the infinite process P with the skeleton tied to the object A has an <u>orientated evolution</u> or is orientated relative to the object A if

$$\lim_{n \to \infty} H_{i_n}(A) = H_\infty(A) \tag{3.7}$$

exists.

The evolution is exact if $H_\infty(A) = 0$.

A process P with the skeleton tied to the object A is <u>stationary</u> relative to the object A if for every two successive elements i' and i'' we have

$$p_{i',i''}(a'|a) = p(a'|a)$$

i.e. if $u_i = u$, $v_i^1 = v$ and $v_i^2 = v^2$ for every $i \in \tilde{N}$.
The morphism $p(a'|a)$ will be called the transition morphism corresponding to the skeleton of the stationary process. In this last case denoting the morphism

$$v^1 \circ u \circ v^2 \in \mathrm{Hom}(A,A)$$

by $p(a'|a)$ we have

$$p_{(n)}(a'|a) = p_{i_0,i_n}(a'|a) = p(a'|a_{n-1}) \circ p(a_{n-1}|a_{n-2}) \circ \ldots$$

$$\ldots \circ p(a_1|a) = (p(a'|a))^n$$

in which we denoted the morphism $p_{i_0,i_n}(a'|a)$ by $p_{(n)}(a'|a)$. We shall call the morphisms v^1, u, v^2 the <u>weights</u> of the stationary process P.

The infinitive process P with the skeleton tied to the object A is <u>homogeneous</u>, relative both to the object A and to the standard morphism $p_0(a) \in \mathrm{Hom}(\{e\}, A)$ if it is stationary relative to A, and if the transition morphism $p(a'|a)$ corresponding to the skeleton of the stationary process P preserves by composition the standard morphism $p_0(a)$ i.e.

$$p(a'|a) \circ p_0(a) = p_0(a').$$

When speaking of a homogeneous process relative to the object A the existence of a standard morphism shall be understood $p_0(a) \in \mathrm{Hom}(\{e\}, A)$ satisfying the last equality.

THEOREM 3.1: Let P be an infinite process stationary relative to the onject A with weights v^1, u, v^2. This process is mixing if, and only if, there is a natural number m_0 and an element $a_0 \in A$ so that for every $a \in A$ there are

$$a_1(a) \in A,\ a_2(a) \in A,\ \dots, a_{m_0-1}(a) \in A$$

so that

$$p(a_1(a)|a) > 0\ ,$$

$$p(a_k(a)|a_{k-1}(a)) > 0\ , \quad (k = 2, \dots, m_0-1)\ , \tag{3.9}$$

$$p(a_0|a_{m_0-1}(a)) > 0$$

where $p(a'|a)$ was denoted by the random morphism $v^1 \circ u \circ v^2 \in \mathrm{Hom}(A,A)$

PROOF: From the well-known theory of Markov chain we have the following result: a stationary Markov chain is mixing if and only if a natural number m_0 and $a_0 \in A$ exist such that

$$p_{(m_0)}(a_0|a) > 0$$

for every $a \in A$. But taking into account the relation (3.4) this is equivalent to saying that

$$\sum_{a_1 \in A} \cdots \sum_{a_{m_0-1} \in A} p(a_0|a_{m_0-1})\, p(a_{m_0-1}|a_{m_0-2}) \cdots p(a_2|a_1)\, p(a_1|a) > 0 \ .$$

But this inequality is satisfied if and only if we have the inequalities (3.9). q.e.d.

COROLLARY 3.1: If the infinite process P stationary relative to the object A is mixing then it is homogeneous relative to the object A and to the standard morphism

$$p_\infty(a') = \lim_{n \to \infty} p_{(n)}(a'|a)$$

where $p_{(n)}(a'|a)$ is the (transition) passage morphism corresponding to the process P during the interval $[\iota_0, \iota_n]$. (The proof is obvious).

<u>Observation</u> 1: Let P be a stationary process relative to the object A with the weights v^1, u, v^2, and let us consider the following square matrix

$$(3.10) \qquad \mathcal{A} = (p(a'|a))_{\substack{a \in A \\ a' \in A}}$$

where $p(a'|a)$ is the random morphism $v^1 \circ u \circ v^2 \in \mathrm{Hom}(A,A)$. In the theory of Markov chains the following assertion is well-known, viz. if number 1 is the single eigen-value with norm equal to 1 of the matrix (3.10) and if this eigen-value has the multiplicity 1, then process P is mixing relative to A.

<u>Observation</u> 2: Obviously, according to the relations (3.4)-(3.7) a mixing process relative to the object A is

orientated relative to this object. The converse is not always true.

Let us now give an example of a process with exact evolution relative to an object A.

THEOREM 3.2: Let P be an infinite process with the skeleton $(u_i)_{i\in\tilde{N}}$ tied to the object $A=\{a_1,a_2,\dots,a_m\}$. Let us suppose that

$$S(u_i) = E(u_i) = X$$

for every $i\in\tilde{N}$. Then the process P has an exact evolution relative to the object A if there exists the standard morphism $p_{i_0}(a)$ where i_0 is the first element of the set $\tilde{N}$ and if for every $x\in X$ we have

$$p_n(a_1|x) = \alpha_x p_{n-1}(a_1|x) + 1 - \alpha_x \qquad (n = 1,2,\dots) \tag{3.11}$$

where $p_n(a|x)$ is the random morphism $v^1_{i_{n+1}} \circ u_{i_n} \in \mathrm{Hom}(X,A)$ and $(\alpha_x)_{x\in X}$ are real numbers so that $0\leq\alpha_x<1$ for every $x\in X$.

PROOF: From (3.11)

$$\sum_{k=2}^{m} p_n(a_k|x) = 1 - p_n(a_1|x) =$$

$$= \alpha_x[1 - p_{n-1}(a_1|x)] = \alpha_x\left(\sum_{k=2}^{m} p_{n-1}(a_k|x)\right)$$

is obtained; by mathematical induction it follows that

$$p_n(a_1|x) = \alpha_x^n p_0(a_1|x) + (1-\alpha_x^n)$$

and also

$$\sum_{k=2}^{m} p_n(a_k|x) = \alpha_x^n \sum_{k=2}^{m} p_0(a_k|x) .$$

Therefore

$$\lim_{n\to\infty} p_n(a_1|x) = 1$$

and

$$\lim_{n\to\infty} \sum_{k=2}^{m} p_n(a_k|x) = 0$$

i.e.

$$\lim_{n\to\infty} p_n(a_k|x) = 0 \qquad (k = 2,\dots, m)$$

for every $x \in X$.

The standard morphism $p_{i_n}(a) \in \mathrm{Hom}(\{e\}, A)$ and the morphism $v_{i_n}^2 \in \mathrm{Hom}(A, X)$ determine the standard morphism $p_{i_n}(x) \in \mathrm{Hom}(\{e\}, A)$ according to the diagram

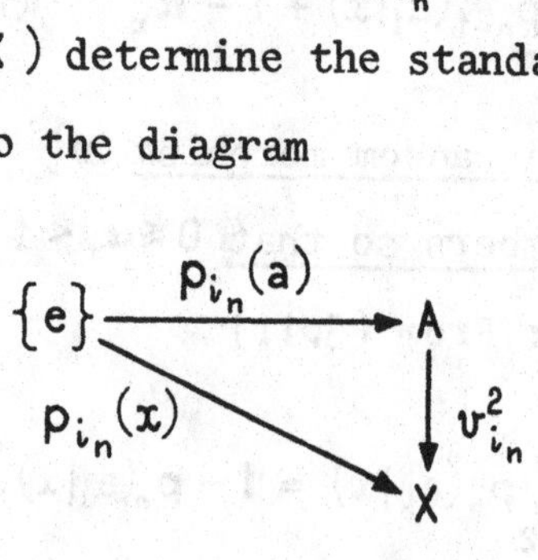

we then have

$$0 \leqslant \lim_{n\to\infty} p_{i_{n+1}}(a_k) = \lim_{n\to\infty} p_n(a_k|x) \circ p_{i_n}(x) = \lim_{n\to\infty} \sum_{x\in X} p_n(a_k|x)\, p_{i_n}(x) \leqslant$$

$$\leq \sum_{x\in X} \lim_{n\to\infty} p_n(a_k|x) = 0 \quad \text{for every} \quad k\,, \quad (k = 2,\ldots,m)\,.$$

Because

$$p_{i_{n+1}}(a_1) + \sum_{k=2}^{m} p_{i_{n+1}}(a_k) = 1\,.$$

Therefore

$$\lim_{n\to\infty} p_{i_{n+1}}(a_1) = 1$$

i.e.

$$\lim_{n\to\infty} H_{i_n}(A) = 0\,.$$ q.e.d.

THEOREM 3.3: Let P be an infinite process with the skeleton $(u_i)_{i\in\tilde{N}}$ tied to the object A. If the standard morphism $p_{i_0}(a)$ exists in which i_0 is the first element of the set $\tilde{N}$ and if for any successive elements i', i'' belonging to $\tilde{N}$ the matrix

$$(p_{i',i''}(a'|a))_{\substack{a\in A\\ a'\in A}} \tag{3.12}$$

is double-stochastic, where $p_{i',i''}(a'|a)$ represents the random morphism $v^1_{i''}\circ u_{i'}\circ v^2_{i'} \in \mathrm{Hom}(A,A)$ then the entropy of the object A increases and process P is orientated relative to A.

PROOF: Let us again put

$$\tilde{N} = \{i_0, i_1, \ldots, i_n, i_{n+1}, \ldots\}\,.$$

Using the well-known inequalities:

$$\sum_{i=1}^{m} q_i \log x_i \leq \log\left(\sum_{i=1}^{m} q_i x_i\right)$$

if

$$q_i \geq 0 \ , \quad \sum_{i=1}^{m} q_i = 1 \ , \quad x_i \geq 1$$

and

$$-\sum_{i=1}^{m} q_i \log q_i \leq -\sum_{i=1}^{m} q_i \log p_i$$

if

$$p_i \geq 0 \ , \quad q_i \geq 0 \ , \quad \sum_{i=1}^{m} p_i = \sum_{i=1}^{m} q_i = 1$$

we obtain

$$H_{i_{k+1}}(A) = -\sum_{a' \in A} p_{i_{k+1}}(a') \log p_{i_{k+1}}(a') =$$

$$= -\sum_{a' \in A} \sum_{a \in A} p_{i_k}(a) p_{i_k, i_{k+1}}(a'|a) \log p_{i_{k+1}}(a') =$$

$$= -\sum_{a \in A} p_{i_k}(a) \log\left(\prod_{a' \in A} [p_{i_{k+1}}(a')]^{p_{i_k, i_{k+1}}(a'|a)}\right) \geq$$

$$\geq -\sum_{a \in A} p_{i_k}(a) \log\left(\sum_{a' \in A} p_{i_k, i_{k+1}}(a'|a) p_{i_{k+1}}(a')\right) \geq$$

$$\geq -\sum_{a \in A} p_{i_k}(a) \log p_{i_k}(a) = H_{i_k}(A) \ .$$

Therefore

$$H_{i_0}(A) \leqslant H_{i_1}(A) \leqslant \ldots \leqslant H_{i_k}(A) \leqslant H_{i_{k+1}}(A) \leqslant \ldots \leqslant \log(\operatorname{card} A)$$

and thus there exists the limit

$$\lim_{k \to \infty} H_{i_k}(A) = H_\infty(A)$$

q.e.d.

Observation 3: This theorem is the famous H theorem of Boltzmann. It is proper to the Markov evolution. Inverse theorems are of special interest to us since we wish to know the conditions implying the decrease of the entropy, i.e.the decrease of the uncertainty and special stress will be laid on this problem at the end of this chapter.

Let us now give examples of processes in FR-categories.

A. The learning process in a learning system

Let $\mathcal{C}$ be a learning system. Let there be $\tilde{N} = N$ where N is the set of natural numbers let

$$P: N \longrightarrow \mathcal{P}(\mathcal{M}(\mathcal{C}))$$

be the infinite process defined in the following manner:

$$P(1) = \{p_1(a), p_1(s,a|a'), p_1(s,a,o|s',a'), 1_{S \times A}, p(a'|s,a)\} \quad (3.16)$$

$$P(n) = \{p_n(s',a'|s,a), p_n(s,a,o|s',a')\circ \tag{3.16}$$
$$\circ p_n(s',a'|s'',a'',o''), p_n(s,a|a'), p(a'|s,a)\} \qquad (n = 2,3,\ldots)$$

where $p_n(s,a\,|\,a')$ is the extension of the morphism $p_n(s|a)$ to the product set $p_n(s,a,o|\,s',a')$ is the extension of the morphism $p_n(o|s,a)$ to the product set and $p(a'|s,a) = \delta_{a,a'}$ is the projection morphism $pr_A \in \mathrm{Hom}(S\times A, A)$. $p_n(s',a'|s,a)$ is given by

$$p_n(s',a'|s,a) = p_n(s',a'|s'',a'',o'')\circ p_{n-1}(s'',a'',o''|s,a) = \tag{3.17}$$
$$= \sum_{o''\in O} p_n(s',a'|s,a,o'')p_{n-1}(o''|s,a) \qquad (n = 2,3,\ldots)$$

and

$$p_n(s|a) = \frac{p_n(s,a)}{p_n(a)} = \frac{p_n(s,a)}{\sum_{s\in S} p_n(s,a)} \tag{3.18}$$

$$\begin{cases} p_n(s,a) = p_n(s,a|s',a',o')\circ p_{n-1}(s',a',o') = \\ = p_n(s,a|s',a',o')\circ p_{n-1}(s',a',o'|s'',a'')\circ p_{n-1}(s'',a'') = \\ = p_n(s,a|s'',a'')\circ p_{n-1}(s'',a'') , \quad (n = 2,3,\ldots) \end{cases} \tag{3.19}$$

$$p_1(s,a) = p_1(a|s)p(s) \; ; \quad p_1(a) = \sum_{s\in S} p_1(s,a) . \tag{3.20}$$

In the process defined above it is possible to

define two skeletons, namely the skeleton $(u^1_n)_{n\in N}$ where

$$u^1_1 = p_1(s,a,o|s',a') ,$$

$$u^1_n = p_n(s,a,o|s',a')\circ p_n(s',a'|s'',a'',o'') , \quad (n = 2,3,\ldots)$$

and respectively the skeleton $(u^2_n)_{n\in N}$ where

$$u^2_1 = 1_{S\times A} , \quad u^2_n = p_n(s',a'|s,a) , \quad (n = 2,3,\ldots) . \tag{3.21}$$

The second skeleton is that of the process P tied to the object A , because for every $n\in N$ there are the morphisms v^1_n and v^2_n from $P(n)$ with

$$S(v^1_n) = S(u_n) , \quad E(v^1_n) = A , \quad S(v^2_n) = A , \quad E(v^2_n) = S(u_n)$$

namely

$$v^1_n = p(a'|s,a) , \quad v^2_n = p_n(s,a|a') . \tag{3.22}$$

The infinite process P defined above may be represented as a diagram, i.e.

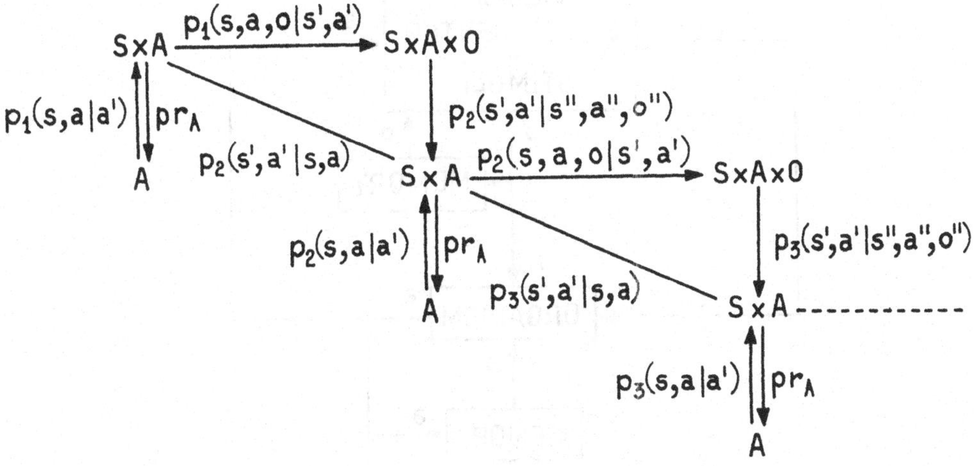

$P(n)$, $(n = 2,3,\dots)$ being formed by the diagram

$$\begin{array}{ccc} S\times A\times O & & \\ \Big\downarrow{\scriptstyle p_n(s',a'|s'',a'',o'')} & & \\ S\times A & \xrightarrow{\;p_n(s,a,o|s',a')\;} & S\times A\times O\ . \\ {\scriptstyle p_n(s,a|a')}\Big\uparrow\Big\downarrow{\scriptstyle pr_A} & & \\ A & & \end{array}$$

The diagrams given above represent the time-development of the process which occurs in the block scheme (customary for engineers) corresponding to the learning systems which may be described as follows:

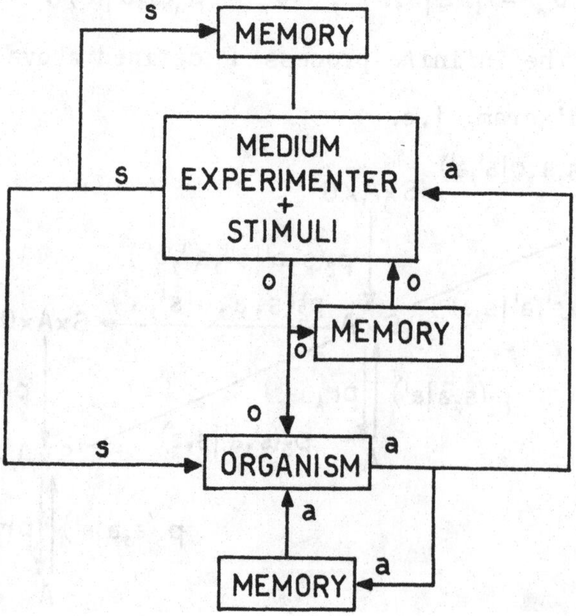

This infinite process presented above gives us exactly the learning process peculiar to a learning process or system. Indeed, in the first experiment we have the stimulus s with the probability $p_1(s)$. In this first experiment the organism gives a response a with the probability $p_1(a|s)$ and this response is followed by a result o with the probability $p_1(o|s,a)$. From these three probabilities it is easy to calculate the probabilities $p_1(s,a), p_1(s,a|a'), p_1(s,a,o|s',a')$. The stimulus s, the response a and the result o, together will modify the probabilities of the possible responses of the organism, the probability of the stimulus s' and the response a' in the following experiment being denoted by $p_2(s',a'|s,a,o)$ and the evolution continues.

When speaking about a given "stimulus" during the first, second, third experiment etc. we understand in fact an ensemble of stimuli denoted by s.

Certainly, according to the proposition 3.1 the skeleton $(u_n^2)_{n\in N}$ defined above, being tied to the object A, induces a Markov chain on A together with the standard morphism

$$p_1(a) = \sum_{s\in S} p_1(s,a).$$

This Markov chain is stationary if, according to (3.2),

$$p_n(s',a'|s,a) = p(s',a'|s,a)$$

$$p_n(s,a|a') = p(s,a|a')$$

for every n

From the theorem 3.1 the following theorem relative to the learning process P occurring in a stationary learning system may be obtained.

THEOREM 3.4: The learning process in a stationary learning system for which $p_n(s|a)$ does not depend on n (and $p(s|a)$ will be written instead of $p_n(s|a)$) is mixing if and only if, a natural number m_0 and a response $\tilde{a}\in A$ exist so that for every response $a\in A$ there exist stimuli $s_k, s'_k, (k=1,2,\dots,m_0)$, a response $a_k (k=1,\dots,m_0-1)$ and a result $o_k (k=1,\dots,m_0)$ in the experiment k, all these depending on a, so that

$$p(s_k,a_k|s'_k,a_{k-1},o_k) > 0$$

(3.23) $$(k=1,\dots,m_0)$$

$$p(o_k|s'_k,a_{k-1}) > 0\ ,\quad p(s'_k|a_{k-1}) > 0$$

for purpose of simplification, denoted as $a_0=a$ and $a_{m_0}=\tilde{a}$ respectively.

PROOF: Let P be the learning process defined by (3.16) in such a way that $p_n(s|a)$ does not depend on n. Denoting the morphism $pr_A\in Hom(S\times A,A)$ by

(3.24) $$p(a|s^*,a^*) = \delta_{a,a^*} = \begin{cases} 1 & \text{if} \quad a = a^* \\ 0 & \text{if} \quad a \neq a^* \end{cases}$$

therefore the morphism $p(a'|a)\in Hom(A,A)$ from the theorem 3.1 will be

$$p(a'|a) = \sum_{s,s',o} p(s',a'|s,a,o)p(o|s,a)p(s|a) \tag{3.25}$$

where to simplify $\sum_{s,s',o}$ is used instead of the triple sum

$$\sum_{s\in S} \sum_{s'\in S} \sum_{o\in O} .$$

For our homogeneous learning system the morphism $p(a'|a)$ defined in the theorem 3.1 is, according to (3.17)-(3.20), (3.24)

$$p(a'|a) = p(a'|s^*,a^*)\circ p(s^*,a^*|s'',a'')\circ p(s'',a''|a) \tag{3.26}$$

where

$$p(s'',a''|a) = p(s''|a'')\,\delta_{a'',a} \tag{3.27}$$

$$\begin{aligned} p(s^*,a^*|s'',a'') &= p(s^*,a^*|\bar{s},\bar{a},\bar{o})\circ p(\bar{s},\bar{a},\bar{o}|s'',a'') = \\ &= \sum_{\bar{o}} p(s^*,a^*|s'',a'',\bar{o})p(\bar{o}|s'',a'') . \end{aligned} \tag{3.28}$$

Therefore, according to (3.24), (3.27) and (3.28), from (3.26) we have

$$\begin{aligned} p(a'|a) &= \sum_{s^*} p(s^*,a'|s'',a'')\circ p(s'',a''|a) = \sum_{s^*,s''} p(s^*,a'|s'',a)p(s''|a) = \\ &= \sum_{s^*,s'',o} p(s^*,a'|s'',a,\bar{o})p(\bar{o}|s'',a)p(s''|a) , \end{aligned}$$

i.e. just the equality (3.25). According to (3.25), obviously (3.23) is a necessary and sufficient condition assuring the validity of (3.9). Finally, it follows from theorem 3.1 that this last condition (3.9) is a necessary and sufficient condition so that the learning process P is mixing. q.e.d.

Therefore there exists a random distribution of the organism's responses which does not depend on the organism's response from the first experiment.

Observation 4: According to the remark 1 it follows also that if the square matrix (3.10) where $p(a'|a)$ is given by (3.25) has the number 1 as the single eigen-value of modulo 1 and if this eigen-value 1 has multiplicity 1, then the learning process P is mixing.

The organism learns if the learning process P defined by (3.16) with the skeleton $(u_n^2)_{n\in N}$ given by (3.21) tied to the object A by (3.22) has an orientated evolution. We shall say that the organism learns completely if the same learning process has an exact evolution.

PROPOSITION 3.2: Let there be a learning system having m possible responses $A = \{a_1, a_2, \dots, a_m\}$. If for every pair $(s,a) \in S\times A$ we have the equality

$$p_n(a_1|s,a) = \alpha(s,a)p_{n-1}(a_1|s,a) + 1 - \alpha(s,a)\,, \quad (n = 2,3,\dots) \tag{3.29}$$

where

$$p_n(a'|s,a) = \sum_{s',o} p(s',a'|s,a,o)p(o|s,a) \tag{3.30}$$

and $\alpha(s,a)$ are real numbers such that $0 \le \alpha(s,a) < 1$ then the organism learns completely.

PROOF: The proposition results immediately from the theorem 3.2 if we take into account the following facts occurring in our case: a) $\tilde{N}$ is the set of natural numbers N b) according to (3.16) the standard morphism $p_1(a) \in P(1)$ exists; c) According to (3.22), in the theorem 3.2, $pr_A \in Hom(S \times A, A)$ is given instead of $v'_{i_{n+1}}$, the product set $S \times A$ instead of X, the pair (s,a) instead of x and the morphism $p_n(s',a' \mid s,a)$ instead of u_{i_n}. With respect to these notations (3.29) is the same as (3.11).

q.e.d.

We shall return to the learning process in the last part of this chapter which is related to the verification of the hypotheses in a predictive system.

B. The process occurring in a random automaton

Let us consider a finite random automaton $\mathcal{C}$ as an FR-category. Again let there be $\tilde{N}$ the set of natural numbers N and let there be the infinite process $P: N \longrightarrow \mathcal{P}(\mathcal{M}(\mathcal{C}))$ defined in the following manner

$$P(1) = \{p_0(a,x), p_1(y|a,x), p_1(a',x'|a,x), 1_{A\times X}, p_1(a,x|a'), p(a'|a,x)\}$$

$$P(n) = \{p_n(y|a,x), p_n(a',x'|a,x), 1_{A\times X}, p_n(a,x|a'), p(a'|a,x)\} \qquad (3.31)$$

$$(n = 2,3,\ldots)$$

where $p_n(a,x|a')$ is the extension of the morphism

$$p_n(x|a) = \frac{p_n(a,x)}{\sum_{x \in X} p_n(a,x)}$$

to the product set and $p(a'|a,x) = \delta_{a',a}$ is the projection morphism $pr_A \in Hom(A \times X, A)$. Of course the category $\mathcal{C}$ being well-equipped all the random correspondences described above are morphisms of the category $\mathcal{C}$.

In the process just given we are able to distinguish a skeleton $(u_n)_{n \in N}$ defined by

$$(3.32) \qquad u_n = p_n(a',x'|a,x) \qquad (n = 1,2,\ldots)$$

which is tied (trivially) to the object $A \times X$ with the morphisms $v_n^1 = v_n^2 = 1_{A \times X}$ and is tied also to the object A with the morphisms

$$(3.33) \quad v_n^1 = p(a'|a,x) \ , \quad v_n^2 = p_n(a,x|a') \ , \quad (n = 1,2,\ldots) \ .$$

The infinite process P defined above may be represented by the following diagram

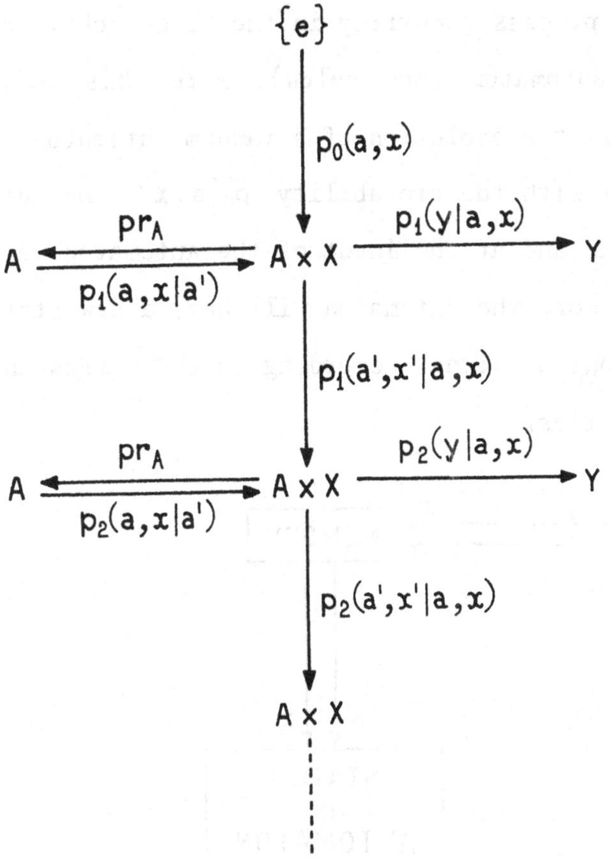

P(n) being represented by the diagram

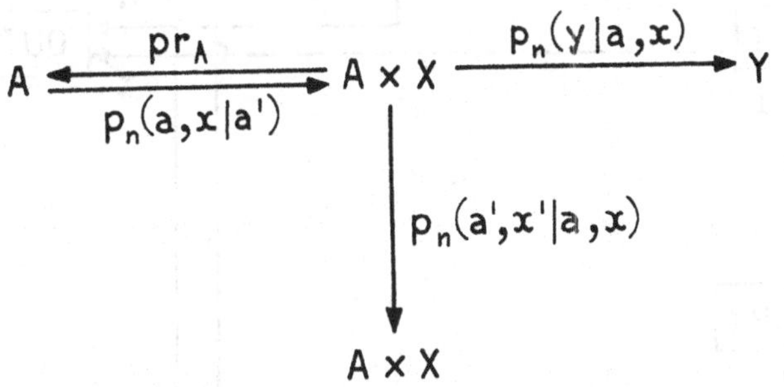

The diagrams given above represent the time-development of the process occurring in the block-scheme corresponding to a random automaton (see below). Here this infinite process represents the evolution of a random automaton. Thus at the initial moment with the probability $p_0(a,x)$ the automaton is in the state a and at the input of the automaton we have the signal x. Furthermore the automaton will have a new state and another input and output signal according to the corresponding transition probabilities.

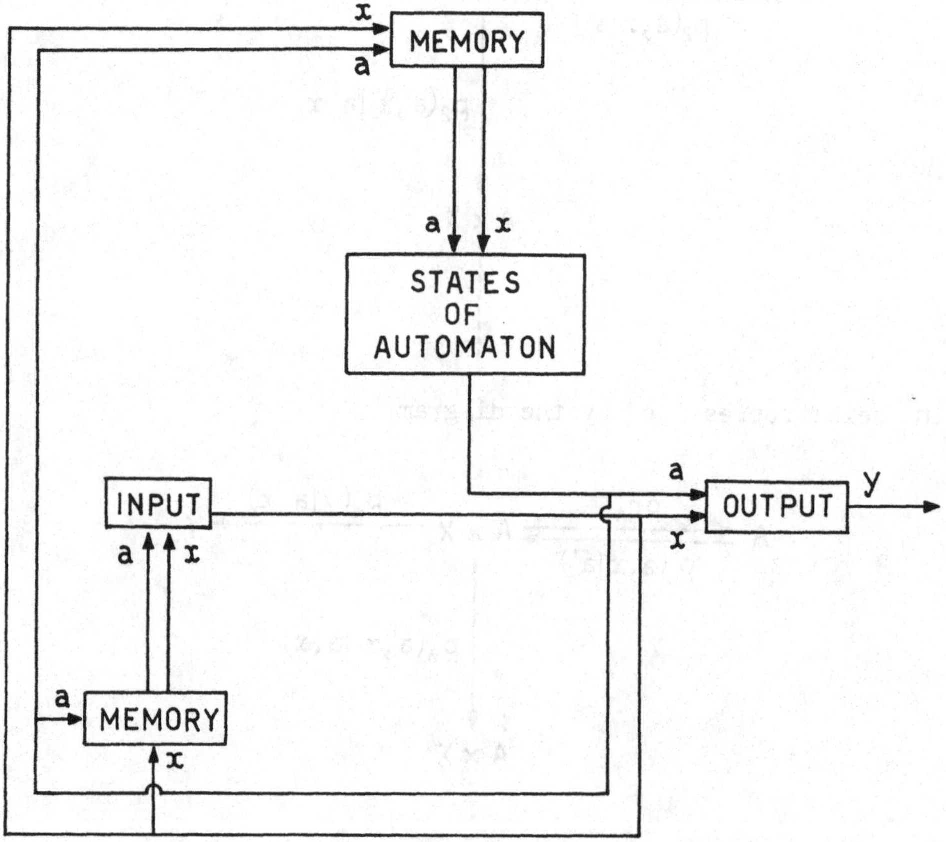

Obviously, according to the property 3.1 the skeleton $(u_n)_{n\in N}$ defined by (3.39) being tied to the object A with the morphism (3.33) it induces on the set A a Markov chain for which the transition probabilities from one state at the moment n to another state at the next moment are given by

$$p_{n,n+1}(a'|a) = p(a'|a'',x'')\circ p_n(a'',x''|a^*,x^*)\circ p_n(a^*,x^*|a) \,. \tag{3.34}$$

We shall say that a random automaton is stationary at the transition if $p_n(a',x'|a,x)$ is independent of n. Similarly a random automaton is stationary at the output if $p_n(y|$ $|a,x)$ does not depend on n. Obviously a random automaton is stationary if and only if it is stationary both at the transition and at the output. The following may be easily obtained from 3.1.

THEOREM 3.5: The process occuring in a random automaton stationary at the transition and for which $p_n(x|a)$ does not depend on n ($p(x|a)$ will be written instead of $p_n(x|a)$) is mixing if and only if a natural number m_0 and an automaton's state $\tilde{a}$ exist so that for every state a input signals x'_k x''_k and automaton's states a_k exist at the moment $k(k=1,\dots,m_0-1)$ all these depending on a, so that

$$p(x'_k|a_{k-1}) > 0\,, \quad p(a_k,x''_k|a_{k-1},x'_k) > 0\,, \quad (k=1,2,\dots,m_0) \tag{3.35}$$

where, for the purpose of simplification the writing $a_0 = a$ and $a_{m_0} = \tilde{a}$ were denoted.

Observation 5: In a random automaton satisfying the condition (3.42) regardless of the initial state of the automaton will occur an asymptotic random stabilization of the states will occur so that a random distribution of the automaton's states will exist which is independent of the initial state.

C. The process occurring in a two-person game

Let us consider as FR-category a two-person game $\mathcal{C}$ defined above. Let there be

$$\tilde{N} = \left\{1, \frac{3}{2}, 2, \frac{5}{2}, 3, \frac{7}{2}, 4, \ldots, \frac{2n-1}{2}, n\right\}$$

where n represents the number of the actions performed by each player, i.e. the duration of the game, and let us consider the finite process

$$P : \tilde{N} \longrightarrow \mathcal{P}(\mathcal{M}(\mathcal{C}))$$

defined by

$$P(k) = \{p_k(x_1, y_1, \ldots, x_{k-1}, y_{k-1}, x_k, y_k | x'_1, y'_1, \ldots, x'_{k-1}, y'_{k-1}, x'_k)\}, (k = 1, 2, \ldots, n)$$

$$P\left(\frac{2l-1}{2}\right) = \left\{p_l(x_1, y_1, \ldots, x_{l-1}, y_{l-1}, x_l | x'_1, y'_1, \ldots, x'_{l-1}, y'_{l-1})\right\}, \quad (l = 2, 3, \ldots, n) \tag{3.36}$$

where

$$p_k(x_1, y_1, \ldots, x_{k-1}, y_{k-1}, x_k, y_k | x'_1, y'_1, \ldots, x'_{k-1}, y'_{k-1}, x'_k)$$

represents the extension of the morphism $p_k(x_k | x_1, y_1, \ldots, x_{k-1}, y_{k-1}, x_k)$

to the product set and

$$p_\ell(x_1, y_1, ..., x_{\ell-1}, y_{\ell-1}, x_\ell | x_1', y_1', ..., x_{\ell-1}', y_{\ell-1}')$$

represents the extension of the morphism $p_\ell(x_\ell | x_1, y_1, ..., x_{\ell-1}, y_{\ell-1})$
to the product set. This process itself represents a skeleton,
namely the skeleton $(u_i)_{i \in \tilde{N}}$ in which the single morphism belonging to the set $P(i)$, was denoted by u_i for every $i \in \tilde{N}$ for the
purpose of simplification. The process defined above may be represented by the diagram given on this page , where $X_i = X$ and
$Y_i = Y$ for every $i, (i = 1, ..., n)$. This is obviously a finite process.

$$
\begin{array}{c}
X_1 \\
\downarrow u_1 \\
X_1 \times Y_1 \xrightarrow{\ u_{3/2}\ } X_1 \times Y_1 \times X_2 \\
\downarrow u_2 \\
\prod_{i=1}^{2}(X_i \times Y_i) \dashrightarrow \\
\qquad \prod_{i=1}^{n-1}(X_i \times Y_i) \xrightarrow{\ u_{\frac{2n-1}{2}}\ } \prod_{i=1}^{n-1}(X_i \times Y_i) \times X_n \\
\qquad\qquad\qquad \downarrow u_n \\
\qquad\qquad\qquad \prod_{i=1}^{n}(X_i \times Y_i)
\end{array}
$$

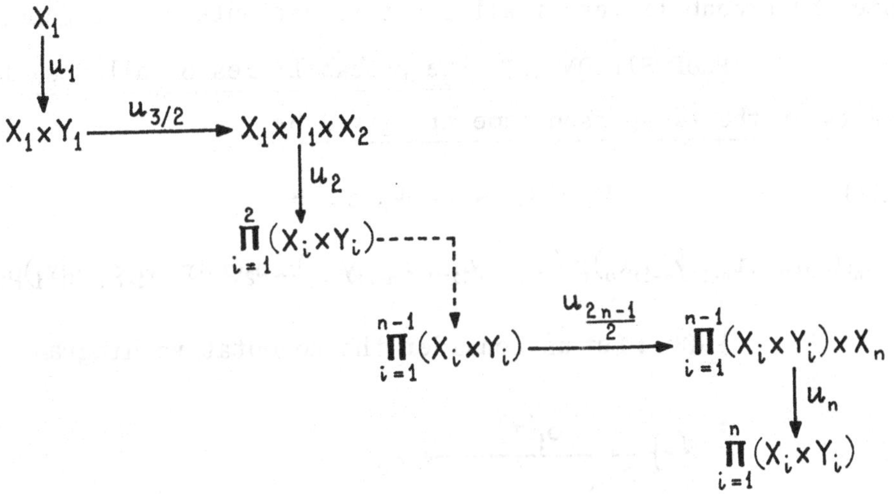

This diagram represents the time-development of
the finite process occurring in a two-person game whose block-
scheme is

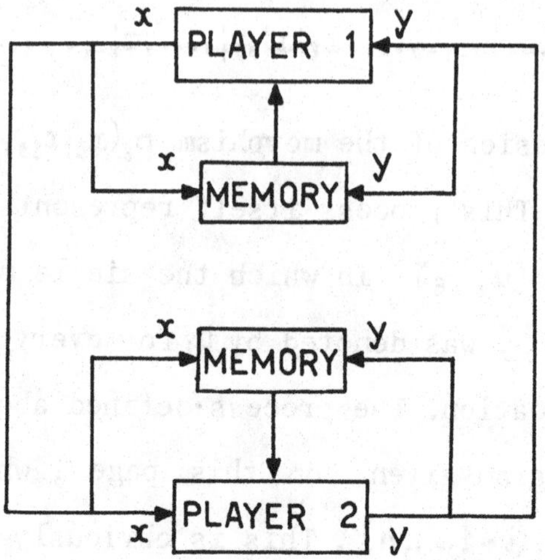

Using the diagram given above it is possible to calculate the probabilities of all possible variants of the game:

PROPOSITION 3.3: <u>The probabilities of all possible variants of the two-person game are given by</u>

(3.37) $$p_{(n)}(x_1, y_1, \dots, x_n, y_n) =$$

$$= p_n(y_n | x_1, y_1, \dots, x_{n-1}, y_{n-1}, x_n)\, p_n(x_n | x_1, y_1, \dots, x_{n-1}, y_{n-1}) \dots p_2(x_2 | x_1, y_1)\, p_1(y_1 | x_1)\, p_1(x_1).$$

PROOF: Let us consider the commutative diagram

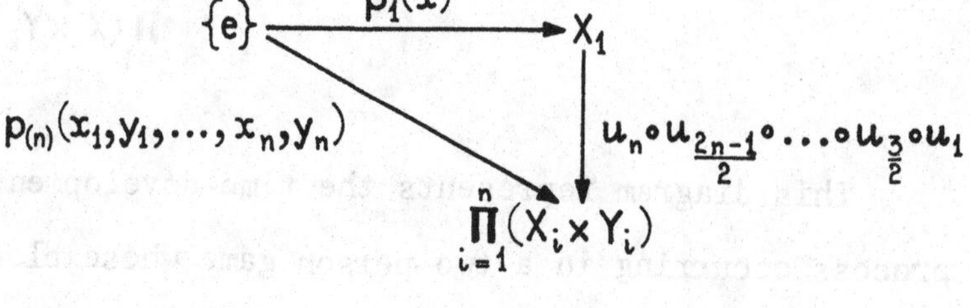

where $X_i = X$, $Y_i = Y$, $(i = 1,2,\dots,n)$. We have

$$P_{(n)}(x_1,y_1,\dots,x_n,y_n) = u_n \circ u_{\frac{2n-1}{2}} \circ \dots \circ u_{\frac{3}{2}} \circ u_1 \circ p_1(x)$$

and replacing the explicit expressions of u_i, $(i \in \tilde{N})$ we obtain (3.37) q.e.d.

D. The process occurring in a predictive system.

Let us consider a predictive system defined in chapter 2.

Let

$$p_k(h|d_{i_1},\dots,d_{i_k})$$

be the reciprocal morphism of the morphism

$$p(d_{i_1},\dots,d_{i_k}|h)$$

if the standard morphism $p_0(h)$ is given, i.e.

$$p_k(h|d_{i_1},\dots,d_{i_k}) = \frac{p_0(h)p(d_{i_1},\dots,d_{i_k}|h)}{\sum_{\bar{h}\in\tilde{H}} p_0(\bar{h})p(d_{i_1},\dots,d_{i_k}|\bar{h})} \qquad (3.38)$$

where $p(d_{i_1},\dots,d_{i_k}|h)$ is the Bayes morphism corresponding to the succession of random morphisms

$$(p(d_{i_\ell}|h,d_{i_1},d_{i_2},\dots,d_{i_{\ell-1}}))_{1\le\ell\le k}$$

namely

$$p(d_{i_1},\ldots,d_{i_k}|h) = \prod_{\ell=1}^{k} p(d_{i_\ell}|h,d_{i_1},\ldots,d_{i_{\ell-1}}) .$$

Thus from (3.38) and (3.39) we obtain

$$p_k(h|d_{i_1},\ldots,d_{i_k}) = \frac{p_0(h)\prod_{\ell=1}^{k} p(d_{i_\ell}|h,d_{i_1},\ldots,d_{i_{\ell-1}})}{\sum_{\bar{h}\in\tilde{H}} p_0(\bar{h})\prod_{\ell=1}^{k} p(d_{i_\ell}|\bar{h},d_{i_1},\ldots,d_{i_{\ell-1}})} . \tag{3.40}$$

Therefore we have the following diagram

$$\begin{array}{ccc} \{e\} & & \\ \downarrow{\scriptstyle p_0(h)} & & \\ \tilde{H} & \xrightarrow{\;p(d_{i_1},\ldots,d_{i_k}|h)\;} & D^k \\ & & \downarrow{\scriptstyle p_k(h|d_{i_1},\ldots,d_{i_k})} \\ & & \tilde{H} \end{array}$$

We now have

$$p_k(h|d_{i_1},\ldots,d_{i_{k-1}},d_{i_k}) = \frac{p_{k-1}(h|d_{i_1},\ldots,d_{i_{k-1}})\,p(d_{i_k}|h,d_{i_1},\ldots,d_{i_{k-1}})}{\sum_{\bar{h}\in\tilde{H}} p_{k-1}(\bar{h}|d_{i_1},\ldots,d_{i_{k-1}})\,p(d_{i_k}|\bar{h},d_{i_1},\ldots,d_{i_{k-1}})} =$$

$$= \frac{p_k(h,d_{i_k}|d_{i_1},\dots,d_{i_{k-1}})}{\sum_{\bar{h}\in\tilde{H}} p_k(h,d_{i_k}|d_{i_1},\dots,d_{i_{k-1}})} = \frac{p_k(h,d_{i_k}|d_{i_1},\dots,d_{i_{k-1}})}{p_k(d_{i_k}|d_{i_1},\dots,d_{i_{k-1}})} .$$

Obviously

$$p(h|d_{i_1},\dots,d_{i_{k-1}}) = \sum_{i_k=1}^{n} p(h,d_{i_k}|d_{i_1},\dots,d_{i_{k-1}}) =$$

$$= \sum_{i_k=1}^{n} p(h|d_{i_1},\dots,d_{i_{k-1}},d_{i_k})p(d_{i_k}|d_{i_1},\dots,d_{i_{k-1}})$$

where $n = \mathrm{card}\, D$, i.e. the number of all possible outcomes.

Let us consider the following entropies

$$H^{(k-1)}_{d_{i_1},\dots,d_{i_{k-1}}}(\tilde{H}) = -\sum_{h\in\tilde{H}} p(h|d_{i_1},\dots,d_{i_{k-1}})\log p(h|d_{i_1},\dots,d_{i_{k-1}})$$

$$H^{(k)}_{d_{i_1},\dots,d_{i_{k-1}}}(\tilde{H}|d_{i_k}) = -\sum_{h\in\tilde{H}} p(h|d_{i_1},\dots,d_{i_{k-1}},d_{i_k})\log p(h|d_{i_1},\dots,d_{i_{k-1}},d_{i_k})$$

$$H^{(k)}_{d_{i_1},\dots,d_{i_{k-1}}}(\tilde{H}|D) = \sum_{i_k=1}^{n} H^{(k)}_{d_{i_1},\dots,d_{i_{k-1}}}(\tilde{H}|d_{i_k})p(d_{i_k}|d_{i_1},\dots,d_{i_{k-1}}) . \tag{3.41}$$

From the fundamental well-known inequality between the entropy and the conditional entropy we immediately have

$$H^{(k)}_{d_{i_1},\dots,d_{i_{k-1}}}(\tilde{H}|D) \leq H^{(k-1)}_{d_{i_1},\dots,d_{i_{k-1}}}(\tilde{H}) , \tag{3.42}$$

which is a form of <u>inverse H-theorem</u> specific to any kind of learning process. Thus the uncertainty on the set of possible hypotheses may only decrease as a consequence of the successive experiments.

Let us consider the particular case

$$(3.43)\quad p(d_{i_\ell}|h,d_{i_1},\dots,d_{i_{\ell-1}}) = p(d_{i_\ell}|h)\,,\quad (\ell = 1,\dots,k)\,.$$

The corresponding Bayes morphism will be

$$(3.44)\quad p(d_{i_1},d_{i_2},\dots,d_{i_k}|h) = p(d_{i_1}|h)p(d_{i_2}|h)\dots p(d_{i_k}|h)$$

and therefore

$$(3.45)\qquad p_k(h|d_{i_1},\dots,d_{i_k}) = \frac{p_0(h)\prod_{\ell=1}^{k} p(d_{i_\ell}|h)}{\sum_{\bar h\in \tilde H} p_0(\bar h)\prod_{\ell=1}^{k} p(d_{i_\ell}|\bar h)}\,.$$

Let us suppose $p_0(h)\neq 0$ whichever be $h\in\tilde H$ because otherwise for such $h^*\in\tilde H$ for which $p_0(h^*)=0$ we have

$$p_k(h^*|d_{i_1},\dots,d_{i_k}) = 0$$

whichever be $d_{i_\ell}\in D,(\ell=1,2,\dots,k)$ which is equivalent to ignoring this hypothesis h^*. At the same time it should be notices that

$$p_{\ell_0}(h|d_{i_1},\dots,d_{i_{\ell_0}}) = 0$$

implies

$$p_\ell(h|d_{i_1},\dots,d_{i_\ell}) = 0$$

for every $\ell > \ell_0$. We shall denote by R the set of hypotheses $h \in \tilde{H}$ for which there exists ℓ_0 such that

$$p_{\ell_0}(h|d_{i_1},\ldots,d_{i_{\ell_0}}) = 0$$

R will be called the set of <u>refutable (in finite time)</u> hypotheses.

The expression (3.45) may be written also in the following form:

$$p_k(h|d_{i_1},\ldots,d_{i_k}) = \frac{p_{k-1}(h|d_{i_1},\ldots,d_{i_{k-1}})p(d_{i_k}|h)}{\sum\limits_{\bar{h}\in\tilde{H}} p_{k-1}(\bar{h}|d_{i_1},\ldots,d_{i_{k-1}})p(d_{i_k}|\bar{h})} . \tag{3.46}$$

Here

$$p_{k-1}(h|d_{i_1},\ldots,d_{i_{k-1}})$$

represents the a priori probability (or credibility) of the hypothesis h before the k-th experiment whereas

$$p_k(h|d_{i_1},\ldots,d_{i_k})$$

may be considered as a posteriori probability (or credibility) of the same hypothesis after the k-th experiment into which the outcome d_{i_k} was obtained. It is of prime importance to underline the essential fact that the Bayes dependence of the "present"

$$p_k(h|d_{i_1},\ldots,d_{i_k})$$

on the "past", i.e.

$$p_{k-1}(h|d_{i_1},\dots,d_{i_{k-1}})$$

is not linear, therefore the process is not a Markov chain. It is interesting to note that the Markov chain is connected to the H-theorem wheas the Bayes chain is connected to the inverse H-theorem.

Let k_i be the absolute frequency of the outcome d_i in the succession

$$B = \{d_{i_1}, d_{i_2}, \dots, d_{i_k}\}$$

obtained in successive experiments. B will be called the "body of evidence" according to the terminology of Satosi Watanabe who has made a noteworthy contribution in this field.

Let us denote by $\alpha_i^{(k)}$ the relative frequency of the outcome d_i in the successive outcomes, i.e.

(3.47) $$\alpha_i^{(k)} = \frac{k_i}{k} .$$

Obviously

$$\sum_{i=1}^{n} \alpha_i^{(k)} = 1 \quad \text{or} \quad \sum_{i=1}^{n} k_i = k .$$

From (3.45) we obtain

(3.48) $$p_k(h|B) = p_k(h|d_{i_1},\dots,d_{i_k}) = \frac{p_0(h)G_B^{(k)}(h)}{\sum_{\bar{h}\in\tilde{H}} p_0(\bar{h})G_B^{(k)}(\bar{h})}$$

where

$$G_B^{(k)}(h) = \left(\prod_{i=1}^{n} [p(d_i|h)]^{\alpha_i^{(k)}} \right)^k . \qquad (3.49)$$

It is easy to see that $p_k(h|B)$ is proportional both to theextraevidential factor $p_0(h)$ and to the purely evidential factor $G_B^{(k)}(h)$. The degree of conformability of the hypotheses h in the presence of the body of evidence B may be considered the expression

$$\Gamma_B^{(k)}(h) = \frac{-\sum_{i=1}^{n} \alpha_i^{(k)} \log \alpha_i^{(k)}}{-\sum_{i=1}^{n} \alpha_i^{(k)} \log p(d_i|h)} . \qquad (3.50)$$

Obviously

$$0 \leqslant \Gamma_B^{(k)}(h) \leqslant 1$$

the right equality being true if and only if

$$p(d_i|h) = \alpha_i^{(k)} , \quad (i = 1,\dots,n)$$

i.e. if empirical and predictive probabilities are perfectly matched. There is an immediate connexion between $G_B^{(h)}(h)$ and $\Gamma_B^{(k)}(h)$ namely

$$G_B^{(k)}(h) = \exp\left(-\frac{-k \sum_{i=1}^{n} \alpha_i^{(k)} \log \alpha_i^{(k)}}{\Gamma_B^{(k)}(h)} \right) .$$

Let us put

$$F_B^{(k)}(h) = \frac{(A_B^{(k)}(h))^k}{\sum_{\bar{h}\in\tilde{H}} (A_B^{(k)}(\bar{h}))^k} = \frac{G_B^{(k)}(h)}{\sum_{\bar{h}\in\tilde{H}} G_B^{(k)}(\bar{h})}$$

where

$$A_B^{(k)}(h) = \prod_{i=1}^{n} (p(d_i|h))^{\alpha_i^{(k)}} .$$

From (3.48) we obtain

$$p_k(h|B) = \frac{p_0(h)F_B^{(k)}(h)}{\sum_{\bar{h}\in\tilde{H}} p_0(\bar{h})F_B^{(k)}(\bar{h})} .$$

Let us suppose the existence of

$$\lim_{k\to\infty} \alpha_i^{(k)} = \gamma_i$$

which corresponds to the case in which the experiments are independent and the outcomes occur with the respective probabilities γ_i . Then we have

$$A_B(h) = \lim_{k\to\infty} A_B^{(k)}(h) = \prod_{i=1}^{n} (p(d_i|h))^{\gamma_i}$$

and

(3.51)
$$\lim_{k\to\infty} F_B^{(k)}(h) = \lim_{k\to\infty} \frac{(A_B(h))^k}{\sum_{\bar{h}\in\tilde{H}} (A_B(\bar{h}))^k} .$$

Because

$$0 \leq A_B(h) \leq 1$$

we have

$$\lim_{k \to \infty} (A_B(h))^k = 0$$

except when the index i exists so that

$$p(d_i|h) = \gamma_i = 1 .$$

Therefore the limit (3.51) can have only two possible values namely o or $1/m$, where m is the number of hipotheses h for which

$$A_B(h) = \max_{\bar{h} \in \tilde{H}} A_B(\bar{h}) . \qquad (3.52)$$

Thus

$$\lim_{k \to \infty} F_B^{(k)}(h) = \frac{1}{m} \qquad (3.53)$$

if the equality (3.52) holds and

$$\lim_{k \to \infty} F_B^{(k)}(h) = 0 \qquad (3.54)$$

if

$$A_B(h) \neq \max_{\bar{h} \in \tilde{H}} A_B(\bar{h}) . \qquad (3.55)$$

Therefore we have

$$\Gamma_B(h) = \lim_{k \to \infty} \Gamma_B^{(k)}(h) = \frac{-\sum_{i=1}^{n} \gamma_i \log \gamma_i}{-\log A_B(h)} . \qquad (3.56)$$

The condition (3.52) may be written in the following form

$$\Gamma_B(h) = \max_{\bar{h} \in \tilde{H}} \Gamma_B(\bar{h}) \tag{3.57}$$

and this means that

$$\lim_{k \to \infty} p_k(h|B) = \frac{p_0(h)}{\sum_{\bar{h} \in K} p_0(h)} \tag{3.58}$$

for $h \in K$ where $K \subset \tilde{H}$ is the set of all hypotheses satisfying (3.57) and

$$\lim_{k \to \infty} p_k(h|B) = 0$$

for $h \notin K$. Thus

$$K = \{h | A_B(h) = \max_{\bar{h} \in \tilde{H}} A_B(\bar{h})\}.$$

We also have

$$R = \{h | A_B(h) = 0\}$$

and the set

$$M = \{h | A_B(h) \neq 0, A_B(h) \neq \max_{\bar{h} \in \tilde{H}} A_B(\bar{h})\} = \tilde{H} - (R \cup K).$$

A classification of all available hypotheses is thereby obtained, so that: R is the set of refutable (in finite time) hypotheses, M is the set of asymptotical refutable hypotheses and K is the class of true (or valid) hypotheses.

Let us consider the random morphisms

$$\bar{p}_k(h|B) = \frac{p_0(h)\bar{F}_B^{(k)}(h)}{\sum_{\bar{h}\in\tilde{H}} p_0(\bar{h})\bar{F}_B^{(k)}(\bar{h})}$$

where

$$\bar{F}_B^{(k)}(h) = (A_B(h))^k .$$

Defining the entropy

$$\bar{H}_B^{(k)} = -\sum_{h\in\tilde{H}} p_k(h|B)\log p_k(h|B)$$

we have the following theorem:

THEOREM 3.6 : <u>There exists</u> k_0 <u>so that for every</u> $k > k_0$ <u>we have</u>

$$\bar{H}_B^{(k-1)} \geqslant \bar{H}_B^{(k)} .$$

PROOF: There is no reason to believe in the existence of changes with respect to the monotonicity of the entropy given above if we pass from discrete time to the continuous one. Thus throughout this proof we shall consider k to be a continuous real variable.

We intend to show that

$$\frac{d\bar{H}_B^{(k)}}{dk} \leqslant 0 .$$

For $h \in R$ we have $A_B(h) = 0$ therefore $\bar{p}_k(h|B) = 0$ for $k \geq 1$. Let us take $h_0 \in K$ and let

$$\beta(h) = \frac{p_0(h)}{p_0(h_0)} > 0$$

$$1 \geq \theta(h) = \frac{A_B(h)}{A_B(h_0)} > 0 .$$

Obviously $\theta(h) = 1$ if $h \in K$. We have $\theta(h) < 1$ if $h \in M$.

$$\bar{H}_B^{(k)} = \frac{\zeta^{(k)}}{\xi^{(k)}}$$

where

$$\xi^{(k)} = \sum_{h \in \tilde{H}} \beta(h)\theta^k(h)$$

$$\zeta^{(k)} = \sum_{h \in M \cup K} \beta(h)\theta^k(h) \log\Big(\sum_{\bar{h} \in M \cup K} \beta(\bar{h})\theta^k(\bar{h})\Big) - \sum_{h \in M \cup K} \beta(h)\theta^k(h) \log(\beta(h)\theta^k(h)) .$$

The derivative is

$$\frac{d\bar{H}_B^{(k)}}{dk} = \frac{1}{(\xi^{(k)})^2}\left(\xi^{(k)}\frac{d\zeta^{(k)}}{dk} - \frac{d\xi^{(k)}}{dk}\zeta^{(k)}\right) \tag{3.59}$$

i.e.

$$\frac{d\bar{H}_B^{(k)}}{dk} = \tag{3.60}$$

$$= \frac{1}{(\xi^{(k)})^2} \sum_{h\in MUK} \sum_{\bar{h}\in MUK} \beta(h)\theta^k(h)\beta(\bar{h})\theta^k(\bar{h}) \log[\beta(h)\theta^k(h)] \log\frac{\theta(\bar{h})}{\theta(h)} .$$

If we add (3.59) and (3.60) replacing h by $\bar{h}$ and $\bar{h}$ by h we obtain

$$2(\xi^{(k)})^2 \frac{d\bar{H}_B^{(k)}}{dk} = \tag{3.61}$$

$$= \sum_{h\in MUK} \sum_{\bar{h}\in MUK} \beta(h)\theta^k(h)\beta(\bar{h})\theta^k(\bar{h}) \log\frac{\beta(h)\theta^k(h)}{\beta(\bar{h})\theta^k(\bar{h})} \log\frac{\theta(\bar{h})}{\theta(h)} .$$

If

$$\log[\beta(h)/\beta(\bar{h})]$$

and

$$\log[\theta(h)/\theta(\bar{h})]$$

have the same sign then

$$\log\frac{\beta(h)\theta^k(h)}{\beta(\bar{h})\theta^k(\bar{h})} \log\frac{\theta(\bar{h})}{\theta(h)} < 0 . \tag{3.62}$$

If

$$\log(\beta(h)/\beta(\bar{h}))$$

and

$$\log(\theta(h)/\theta(\bar{h}))$$

have opposite signs, we again have (3.62) for k of the form

(3.63) $$k > -\frac{\log(\beta(h)/\beta(\bar{h}))}{\log(\theta(h)/\theta(\bar{h}))} .$$

If

$$\log(\beta(h)/\beta(\bar{h})) = 0$$

$$\log(\theta(h)/\theta(\bar{h})) \neq 0 ,$$

(3.62) is true for any k . If

$$\log(\theta(h)/\theta(\bar{h})) = 0$$

then

(3.64) $$\log\frac{\beta(h)\theta^k(h)}{\beta(\bar{h})\theta^k(\bar{h})}\log\frac{\theta(\bar{h})}{\theta(h)} = 0$$

whatever the value of

$$\log(\beta(h)/\beta(\bar{h})) .$$

From (3.61)-(3.63) we have

$$\frac{d\bar{H}_B^{(k)}}{dk} \leqslant \theta$$

for k sufficiently large such that (3.73) be satisfied by those pairs $(h,\bar{h})$ which give different values for θ . A finite lower bound for these k must exist because the total number of all available hypotheses is finite. q.e.d.

Observation 6: The progress of evaluation of available hypotheses in a predictive system and the theorem 3.6 are both due to S. Watanabe (see [17]).

Chapter 4

THE REDUCTION OF ONE RANDOM MORPHISM TO AN ε-DETERMINISTIC ONE

A deterministic morphism is a particular case of a random morphism. On the other hand a bijective deterministic morphism (bijection) with the source X and the ending X generates a random morphism with the source and respective ending equal to $X/\mathbf{R}$ where $\mathbf{R}$ is the equivalence relation. Therefore if we consider a bijection of the set X on itself (i.e. a special case of deterministic morphism) and we pass to a poorer set (the set of equivalence class with respect to a given equivalence relation on X i.e. a partition of X) then a random morphism is obtained.

The converse problem, that of starting from a given morphism in order to attach a deterministic morphism to it, is much more difficult. It will nevertheless be shown here that in some circumstances, by starting from a given random morphism we are able to arrive at an ε-deterministic morphism (i.e. to an almost deterministic morphism, or to a deterministic morphism with an error smaller than ε) if we take into account the existence of some richer sets which will constitute the new source and the new ending, i.e. the new source will be the product of the source of the initial random morphism with itself, and the new ending will be the product of the ending of the initial random morphism by itself.

Let us consider an FR-category $\mathcal{C}$ well-equipped and let there be the morphisms

$$p(x):\{e\} \longrightarrow X \; ; \quad p(x'|x):X \longrightarrow X \; ; \quad p(y|x):X \longrightarrow Y$$

together with the derived morphisms

$$p(x,y):\{e\} \longrightarrow X \times Y \, ,$$
$$p(y):\{e\} \longrightarrow Y$$
$$p(x|y):Y \longrightarrow X$$

where

$$p(x,y) = p(y|x)p(x) \, , \quad p(y) = \sum_{x \in X} p(x,y) \tag{4.1}$$
$$p(x|y) = \frac{p(x,y)}{p(y)} \, .$$

The morphism

$$p(y|x):X \longrightarrow Y$$

defines a family of measures $\{\nu_\omega\}_{\omega \in X^I}$ where for every $\omega \in X^I$ (I is the set of integer numbers), ν_ω is a measure on $\mathfrak{F}_Y$ where $\mathfrak{F}_Y$ is the σ -algebra generated by the family of cylinder sets belonging to Y^I . More exactly for

$$\omega = (\dots, x_{-1}, x_0, x_1, \dots, x_n, x_{n+1}, \dots) \in X^I$$

and for

$$[\bar{y}_1, \dots, \bar{y}_n] = \{(\dots, y_{-1}, y_0, y_1, \dots, y_n, \dots) | y_i \in Y, y_1 = \bar{y}_1, \dots, y_n = \bar{y}_n\} \subset Y^I$$

we define

$$\nu_\omega([\bar{y}_1, \dots, \bar{y}_n]) = p(\bar{y}_1|x_1) \dots p(\bar{y}_n|x_n) .$$

Obviously, according to this definition it follows that for every

$$\omega \in [x_1, \dots, x_n] \subset X^I$$

and for every

$$\sigma \in [y_1, \dots, y_n] \subset Y^I$$

we have

$$\nu_\omega(\sigma) = p(y_1|x_1) \dots p(y_n|x_n) . \qquad (4.2)$$

The family of measures $\{\nu_\omega\}_{\omega \in X^I}$ will be called the family of measures generated by the morphism $p(y|x)$.

Let us consider also the Markov measure generated by the morphisms

$$p(x) : \{e\} \longrightarrow X , \quad p(x'|x) : X \longrightarrow X$$

defined on the σ-algebra $\mathfrak{F}_X$. Thus for every cylinder set

$$[x_1, \ldots, x_n] \subset X^I$$

this Markov measure has the following explicit expression

$$(4.3) \quad \mu([x_1, \ldots, x_n]) = p(x_1)p(x_2|x_1)\ldots p(x_n|x_{n-1}) = p(x_1, \ldots, x_n).$$

Assuming that

$$p(x) \circ p(x'|x) = p(x').$$

Then obviously the Markov measure μ is stationary with respect to the shift operator T, where this last operator is usually defined as

$$T: X^I \longrightarrow X^I$$

so that

$$T(\ldots, x_{-2}, x_{-1}, x_0, x_1, x_2, \ldots) = (\ldots, x'_{-2}, x'_{-1}, x'_0, x'_1, x'_2, \ldots)$$

where

$$x'_k = x_{k+1} \quad (k \in I).$$

Clearly, the random morphisms $p(x)$ and $p(x'|x)$ given above completely determine the Markov morphisms of different orders defining just the probabilities of the cylinder sets from $\mathfrak{F}_X$ which completely determine the measure μ on $\mathfrak{F}_X$. Let

$$\prod_{i=1}^{n} X_i \xrightarrow{p(y_1,\ldots,y_n|x_1,\ldots,x_n)} \prod_{i=1}^{n} Y_i \qquad (X_i = X\,,\ Y_i = Y)\,,$$

be the product morphism of order n generated by the random morphism

$$p(y|x): X \longrightarrow Y$$

and defined by the equality

$$p(y_1,\ldots,y_n|x_1,\ldots,x_n) = p(y_1|x_1)\ldots p(y_n|x_n)\,. \qquad (4.4)$$

According to (4.3) and (4.4) for every

$$\omega \in [x_1,\ldots,x_n] \subset X^I$$

and every

$$\sigma \in [y_1,\ldots,y_n] \subset Y^I$$

we have

$$\nu_\omega(\sigma) = p(y_1,\ldots,y_n|x_1,\ldots,x_n)\,. \qquad (4.5)$$

Given both the Markov morphism $p(x_1,\ldots,x_n)$ of order $(n-1)$ generated by the morphisms $p(x) \in \mathrm{Hom}(\{e\},X)$, $p(x'|x) \in \mathrm{Hom}(X,X)$ i.e. the morphism

$$p(x_1,\ldots,x_n) \in \mathrm{Hom}\Big(\{e\}, \prod_{i=1}^{n} X_i\Big)\,, \qquad (X_i = X)$$

and the product morphism $p(y_1,\ldots,y_n|x_1,\ldots x_n)$ of order n generated by the morphism $p(y|x) \in \mathrm{Hom}(X,Y)$ let there be the derived morphisms

$$p(x_1,\dots,x_n,y_1,\dots,y_n)\in \mathrm{Hom}\left(\{e\},\prod_{i=1}^{n}X_i\times\prod_{i=1}^{n}Y_i\right)$$

$$(X_i = X\,,\quad Y_i = Y)$$

$$p(y_1,\dots,y_n)\in \mathrm{Hom}\left(\{e\},\prod_{i=1}^{n}Y_i\right),\quad (Y_i = Y)$$

$$p(x_1,\dots,x_n|y_1,\dots,y_n)\in \mathrm{Hom}\left(\prod_{i=1}^{n}Y_i,\prod_{i=1}^{n}X_i\right),\quad (X_i = X,\ Y_i = Y)$$

defined by

$$p(x_1,\dots,x_n,y_1,\dots,y_n) = p(y_1,\dots,y_n|x_1,\dots,x_n)p(x_1,\dots,x_n) =$$

$$= p(y_1|x_1)\dots p(y_n|x_n)p(x_1,\dots,x_n)$$

(4.6)

$$p(y_1,\dots,y_n) = \sum_{x_1\in X,\dots,x_n\in X} p(x_1,\dots,x_n,y_1,\dots,y_n)$$

$$p(x_1,\dots,x_n|y_1,\dots,y_n) = \frac{p(x_1,\dots,x_n,y_1,\dots,y_n)}{p(y_1,\dots,y_n)}\,.$$

All these morphisms are well-defined for every n therefore the measure λ on $\mathfrak{F}_{X\times Y}$ and the measure η on $\mathfrak{F}_Y$ are completely determined where λ is the measure determined by $p\cdot(x_1,\dots,x_n,y_1,\dots,y_n)$ and η is the measure determined by $p(y_1,\dots,y_n)$ for $(n = 1,2,\dots)$.

We shall say therefore that the morphism

$$p(x'|x) \in \mathrm{Hom}(X,X)$$

is <u>ergodic</u> if the limit

$$\lim_{n \to \infty} \frac{1}{n} \sum_{k=0}^{n-1} p_{(k)}(x'|x) = p(x') \tag{4.7}$$

exists for every $x \in X$ where $p_{(k)}(x'|x)$ represents the k-th power of the morphism $p(x'|x)$, defined by the equality

$$p_{(k)}(x'|x) = p(x'|x_{k-1}) \circ p(x_{k-1}|x_{k-2}) \circ \ldots \circ p(x_1|x) \tag{4.8}$$

It is therefore possible to demonstrate easily that if

$$p(x) \circ p(x'|x) = p(x') \tag{4.9}$$

and if the morphism $p(x'|x)$ is ergodic then the Markov measure μ generated by the two morphisms given above is ergodic with respect to the shift operator (i.e. the shift operator is ergodic with respect to the Markov measure μ).

Let the entropies be written as follows

$$H(p(y|x)) = H_{p(x)}(p(y|x)) = -\sum_{x \in X} p(x) \sum_{y \in Y} p(y|x) \log p(y|x) \tag{4.10}$$

$$H(p(x)) = -\sum_{x \in X} p(x) \log p(x) \tag{4.11}$$

and let

(4.12) $r(p(x)) = H(p(x)) - H_{p(y)}(p(x|y)) = H(p(y)) - H_{p(x)}(p(y|x))$.

Let

(4.13) $$C = \sup_{p(x)} r(p(x)) .$$

As in the usual information theory it is easy to prove the existence of the limit

(4.14) $$\lim_{n\to\infty} \frac{1}{n} H(p_n(x_1, \ldots, x_n)) = H$$

$$\sum_{x_i \in X} p_i(x_1,\ldots,x_i) = p_{i-1}(x_1,\ldots,x_{i-1}); \quad \sum_{x\in X} p_{i+1}(x,x_1,\ldots,x_i) = p_i(x_1,\ldots,x_i)$$

(the number H will be called the entropy of the stationary measure generated by the sequence of the random morphisms $(p_i(x_1,\ldots$ $\ldots,x_i)_{1\le i\le\infty})$.

Let us suppose now that the random morphism $p(y|x)$ is fixed. Then for every pair of morphisms $p(x)$ and $p(x'|x)$ (i.e. for every stationary Markov measure μ_m)there is a corresponding value for

$$R\mu_m = \lim_{n\to\infty} \frac{1}{n}[H(p(y_1,\ldots,y_n)) - H_{p(x_1,\ldots,x_n)}(p(y_1,\ldots,y_n|x_1,\ldots,x_n))] =$$

(4.15) $$= \lim_{n\to\infty} \frac{1}{n}[H(p(y_1,\ldots,y_n)) - nH_{p(x)}(p(y|x))] = H^* - H_{p(x)}(p(y|x)).$$

Given the number

$$C_{ms} = \sup_{\mu_{m_s}} R\mu_m$$

where sup. is taken over all possible stationary Markov measures, i.e. over all possible random morphisms $p(x'|x)$ and $p(x)$ so that

$$p(x'|x)\circ p(x) = p(x')$$

and also the number

$$C_{mx} = \sup_{\mu_{mx}} R\mu_m$$

where sup. is taken on over all possible random morphisms so that

$$p(x'|x)\circ p(x) = p(x')$$

$$\lim_{n\to\infty} \frac{1}{n} \sum_{k=0}^{n-1} p_{(k)}(x'|x) = p(x')$$

i.e. over all possible stationary ergodic Markov measures.

According to the information theory it is easy to prove that (see [13]),

$$C_{mx} = C_{ms} = C . \tag{4.13'}$$

It is, therefore, also possible to prove that the measure λ on $\mathfrak{F}_{X\times Y}$ generated by the sequence of the standard morphisms

$$(p(x_1,y_1,\ldots,x_n,y_n))_{1\leq n<\infty}$$

is ergodic with respect to the shift operator. It follows immediately that the measure

$$\eta(N) = \lambda(X^I \times N), \quad N \in \mathfrak{F}_Y$$

is also stationary and ergodic.

By using the known Feinstein techniques we obtain the following theorem:

THEOREM 4.1: Given the random morphism

$$p(y|x) \in \mathrm{Hom}(X,Y)$$

The natural number $n_0(\varepsilon)$ exists for every $\varepsilon > 0$ so that for every $n \geq n_0(\varepsilon)$ there exist both the set $A_n \subset X^n$ having more that $2^{n(C-\varepsilon)}$ elements where

$$C = \sup_{p(x)} \left\{ \sum_{\substack{x \in X \\ y \in Y}} p(x)p(y|x) \log \frac{p(y|x)}{\sum_{x \in X} p(x)p(y|x)} \right\}, \tag{4.16}$$

and a partition $\pi(Y^n)$ of the product set Y^n such that the restriction to the object A_n of the extension of the product morphism of order n generated by $p(y|x)$, to the partition $\pi(Y^n)$, is an ε -deterministic morphism.

PROOF: According to (4.12) let us observe that the number C defined by the equality (4.16) is just the number C defined by (4.13). According to (4.13') a Markov stationary ergodic measure μ exists such that

$$R_\mu > C - \frac{\varepsilon}{2}$$

The Markov measure μ is generated by the morphisms

$$p(x) \in \mathrm{Hom}(\{e\}, X) \ , \quad p(x'|x) \in \mathrm{Hom}(X, X) \, .$$

Let there be the sequence of Markov morphisms

$$(p(x_1, \dots, x_n))_{1 \le n < \infty}$$

of different orders generated by the morphisms $p(x)$ and $p(x'|x)$. Also let there be the derived morphisms

$$p(x_1, y_1, \dots, x_n, y_n) = p(x_1, \dots, x_n) p(y_1, \dots, y_n | x_1, \dots, x_n)$$

$$p(y_1, \dots, y_n) = \sum_{x_1 \in X, \dots, x_n \in X} p(x_1, y_1, \dots, x_n, y_n)$$

where $p(y_1, \dots, y_n | x_1 \dots x_n)$ represents the product morphism of order n generated by the morphism $p(y|x)$. Let λ be the stationary measure on $\mathfrak{F}_{X \times Y}$ generated by the sequence of morphisms

$$(p(x_1, y_1, \dots, x_n, y_n))_{1 \le n < \infty}$$

and η be the stationary measure on $\mathfrak{F}_Y$ generated by the sequence of morphisms

$$(p(y_1, \dots, y_n))_{1 \le n < \infty}$$

of course, λ and η are ergodic too. Then, from well-known McMillan theorem it follows that the sequences

$$-\frac{1}{n}\log\mu(X_n), \quad -\frac{1}{n}\log\eta(Y_n), \quad -\frac{1}{n}\log\lambda(X_n\times Y_n)$$

converge in probability λ to the mean values H_X, H_Y; and respectively $H_{X\times Y}$ where we put

$$X_n = [x_1,\ldots,x_n], \quad Y_n = [y_1,\ldots,y_n]$$

$$X_n\times Y_n = [(x_1,y_1),\ldots,(x_n,y_n)]$$

and

$$H_X = \lim_{n\to\infty}\frac{1}{n}H(p(x_1,\ldots,x_n)); \quad H_Y = \lim_{n\to\infty}\frac{1}{n}H(p(y_1,\ldots,y_n))$$

$$H_{X\times Y} = \lim_{n\to\infty}\frac{1}{n}H(p(x_1,y_1,\ldots,x_n,y_n)).$$

Obviously

$$\mu(X_n) = \mu([x_1,\ldots,x_n]) = p(x_1,\ldots,x_n)$$

$$\eta(Y_n) = \eta([y_1,\ldots,y_n]) = p(y_1,\ldots,y_n)$$

$$\lambda(X_n\times Y_n) = \lambda([(x_1,y_1),\ldots,(x_n,y_n)]) = p(x_1,y_1,\ldots,x_n,y_n).$$

Therefore the sequence

$$\frac{1}{n}\log\frac{\lambda(X_n\times Y_n)}{\mu(X_n)\eta(Y_n)}$$

converges in the probability λ towards the limit

$$H_X + H_Y - H_{X\times Y} = R_\mu > C - \frac{\varepsilon}{2} .$$

Furthermore

$$\nu_{X_n}(Y_n) = \frac{\lambda(X_n \times Y_n)}{\mu(X_n)} .$$

Then

$$\nu_{X_n}(Y_n) = \nu_{[x_1,\dots,x_n]}([y_1,\dots,y_n]) = \frac{p(x_1,y_1,\dots,x_n,y_n)}{p(x_1,\dots,x_n)} = p(y_1,\dots,y_n|x_1,\dots x_n).$$

For $n_0(\varepsilon)$ sufficiently large, if we take $n>n_0(\varepsilon)$ we have

$$\lambda\left(\frac{1}{n}\log\frac{\nu_{X_n}(Y_n)}{\eta(Y_n)} > C - \frac{\varepsilon}{2}\right) > 1 - \frac{\varepsilon}{2} . \tag{4.17}$$

For every X_n let Y_{X_n} be the union of the cylinder sets

$$Y_n = [y_1,\dots,y_n]$$

for which

$$\frac{1}{n}\log\frac{\nu_{X_n}(Y_n)}{\eta(Y_n)} > C - \frac{\varepsilon}{2} . \tag{4.18}$$

According to (4.17) we have

$$1-\frac{\varepsilon}{2} < \lambda\left(\frac{1}{n}\log\frac{\nu_{X_n}(Y_n)}{\eta(Y_n)} > C - \frac{\varepsilon}{2}\right) = \lambda(\bigcup_{X_n}(X_n \times Y_{X_n})) =$$

(4.19)

$$= \sum_{X_n}\lambda(X_n \times Y_{X_n}) = \sum_{X_n}\mu(X_n)\nu_{X_n}(Y_{X_n}) .$$

If X_n and Y_n satisfy (4.18) we have

(4.20) $$\nu_{X_n}(Y_n) > 2^{n\left(C-\frac{\varepsilon}{2}\right)}\eta(Y_{X_n})$$

i.e.

$$1 \geqslant \nu_{X_n}(Y_{X_n}) > 2^{n\left(C-\frac{\varepsilon}{2}\right)}\eta(Y_{X_n})$$

and thus

(4.21) $$\eta(Y_{X_n}) < 2^{-n\left(C-\frac{\varepsilon}{2}\right)} .$$

It is possible to assume without any loss of generality that

(4.22) $$n_0(\varepsilon) \geqslant \frac{2}{\varepsilon}\log\frac{2}{\varepsilon}$$

and let $X_n^{(1)}$ be the cylinder set

$$X_n = [x_1, \ldots, x_n]$$

so that

$$\nu_{X_n}(Y_{X_n}) > 1 - \varepsilon$$

which according to (4.19) is possible, which can be denoted by

$$Y^{(1)} = Y_{X_n^{(1)}}$$

and let $X_n^{(2)}$ be a cylinder set $X_n = [x_1,\dots,x_n]$ (assuming that such a cylinder set exists) so that

$$\nu_{X_n}(Y_{X_n} - Y^{(1)}) > 1 - \varepsilon$$

and

$$Y^{(2)} = Y_{X_n^{(2)}} - Y^{(1)} .$$

Furthermore, let $X_n^{(3)}$ be a cylinder set X_n if such a cylinder set exists, for which

$$\nu_{X_n}(Y_{X_n} - Y^{(1)} - Y^{(2)}) > 1 - \varepsilon$$

and

$$Y^{(3)} = Y_{X_n^{(3)}} - Y^{(1)} - Y^{(2)} .$$

This construction may be continued only for a finite number of steps. Let

$$X_n^{(1)},\dots,X_n^{(N)}; \quad Y^{(1)},\dots,Y^{(N)}$$

be the sets obtained in this way. According to the construction described above we have

$$Y^{(i)} \cap Y^{(k)} = \emptyset, \quad (i \neq k), \quad (i,k = 1,\dots,N)$$

and

$$(4.23) \qquad \nu_{X_n^{(i)}}(Y^{(i)}) > 1-\varepsilon\,, \quad (i = 1,\dots,N)\,.$$

For every X_n we have

$$(4.24) \qquad \nu_{X_n}(Y_{X_n} - \bigcup_{i=1}^{N} Y^{(i)}) \leq 1-\varepsilon$$

and thus

$$\nu_{X_n}(Y_{X_n}) \leq \nu_{X_n}(\bigcup_{i=1}^{N} Y^{(i)}) + \nu_{X_n}(Y_{X_n} - \bigcup_{i=1}^{N} Y^{(i)}) \leq$$

$$\leq \nu_{X_n}(\bigcup_{i=1}^{N} Y^{(i)}) + 1 - \varepsilon$$

i.e.

$$\sum_{X_n} \mu(X_n)\nu_{X_n}(Y_{X_n}) \leq \eta(\bigcup_{i=1}^{N} Y^{(i)}) + 1 - \varepsilon$$

where the sum is taken over all possible cylinder sets X_n.

According to (4.19) from the last inequality it results

$$(4.25) \qquad \eta(\bigcup_{i=1}^{N} Y^{(i)}) > \frac{\varepsilon}{2}\,.$$

But

$$Y^{(i)} \subset Y_{X_n^{(i)}} \qquad (i = 1,\dots,N)$$

and from (4.21) we have

$$\eta\left(\bigcup_{i=1}^{N} Y^{(i)}\right) \leq \sum_{i=1}^{N} \eta(Y_{X_n^{(i)}}) < N 2^{-n\left(C-\frac{\varepsilon}{2}\right)}. \tag{4.26}$$

The inequalities (4.25) and (4.26) imply

$$N > \frac{\varepsilon}{2} 2^{n\left(C-\frac{\varepsilon}{2}\right)}. \tag{4.27}$$

For $n > n_0(\varepsilon)$ from (4.22) we have

$$n\frac{\varepsilon}{2} > \log\frac{2}{\varepsilon}, \quad \text{i.e.} \quad \frac{\varepsilon}{2} > 2^{-n\frac{\varepsilon}{2}}$$

and according to (4.27) we obtain

$$N > 2^{n(C-\varepsilon)}.$$

Every cylinder set $X_n^{(i)}$, $(1 \leq i \leq N)$ is a set of the form

$$X_n^{(i)} = [x_1^{(i)},\dots,x_n^{(i)}] \subset X^I, \quad (1 \leq i \leq N)$$

and every set $Y^{(i)}$, $(1 \leq i \leq N)$ is, according to the construction given above, a union of the cylinder sets $Y_n = [y_1,\dots,y_n]$ namely

$$Y^{(i)} = \bigcup_{j=1}^{m_i} Y_n^{(j)} = \bigcup_{j=1}^{m_i} ([y_1^{(j)},\dots,y_n^{(j)}]).$$

Let $\mathfrak{x}_n^{(i)}$ be the projection of $X_n^{(i)}$ on the set denoted by

$$X^n = \prod_{k=1}^{n} X_{(k)}$$

where

$$X_{(k)} = X\,, \quad (k = 1,\dots,n)$$

i.e.

$$\bar{\mathfrak{x}}_n^{(i)} = (x_1^{(i)},\dots,x_n^{(i)})$$

and by $\bar{y}^{(i)}$ the projection of $Y^{(i)}$ on the set

$$Y^n = \prod_{k=1}^{n} Y_{(k)}$$

where

$$Y_{(k)} = Y\,, \quad (k = 1,\dots,n)$$

i.e.

$$\bar{y}^{(i)} = \bigcup_{j=1}^{m_i} (y_1^{(j)},\dots,y_n^{(j)}) = \bigcup_{j=1}^{m_i} \bar{y}_n^{(j)} \subset Y^n\,.$$

Of course

$$Y^{(j)} \cap Y^{(k)} = \emptyset\,, \quad (j \neq k)$$

implies

$$\bar{y}^{(j)} \cap \bar{y}^{(k)} = \emptyset\,, \quad (j \neq k)$$

Let us now take

$$A_n = \{\bar{\mathfrak{x}}_n^{(1)},\dots,\bar{\mathfrak{x}}_n^{(N)}\} \subset X^n$$

and consider the partition

$$\pi(Y^n) = \{\bar{y}^{(1)},\dots,\bar{y}^{(N)},\bar{y}^{(N+1)}\}$$

in which

$$\bar{y}^{(N+1)} = Y^n - \bigcup_{i=1}^{N} \bar{y}^{(i)} .$$

Taking into account the definition of the measures $\{\nu_\omega\}_{\omega\in A^I}$ the morphism

$$p(\bar{y}_n|\bar{x}_n) = p(y_1,\dots,y_n|x_1,\dots,x_n) = \nu_{[x_1,\dots,x_n]}([y_1,\dots,y_n])$$

is well-defined where by

$$\bar{x}_n = (x_1,\dots,x_n)\in X^n , \quad \bar{y}_n = (y_1,\dots,y_n)\in Y^n .$$

According to (4.23) $\bar{y}^{(i)}\in\pi(Y^n)$ exists for every $\bar{æ}_n^{(i)}\in A_n$ so that

$$p(\bar{y}^{(i)}|\bar{æ}_n^{(i)}) > 1-\varepsilon$$

the sets $\bar{y}^{(i)},(1\leq i\leq N)$ being pairwise disjoint. This means that the canonical morphism of transition to the partition $\pi(Y^n)$

$$p(\bar{y}|\bar{x})\in \mathrm{Hom}(A_n,\pi(Y^n))$$

obtained from the restriction to A_n of the product morphism

$$p(y_1,\dots,y_n|x_1,\dots,x_n)$$

generated by $p(y|x)$ is ε-deterministic. q.e.d.

The ε-deterministic morphism whose existence has just been proved will be called the ε-deterministic morphism associated with the morphism $p(y|x)$.

Obviously we have the inclusion

$$A_n \subset X^n = \prod_{k=1}^{n} X_{(k)}, \quad (X_{(k)} = X).$$

According to the last theorem the restriction to the set A_n of the canonical morphism of transition to the partition $\pi(Y^n)$ obtained from the product morphism

$$p(y_1, \ldots, y_n | x_1, \ldots, x_n)$$

is ε-deterministic, which in turn poses a question as to when exactly is the sourse A_n of this ε-deterministic morphism sufficiently large. In other words, when is the set A_n sufficiently as large as the whole X^n. The following theorem will now be proved as an answer to this question.

THEOREM 4.2: Given the random morphism

$$p(y|x) \in \mathrm{Hom}(X, Y)$$

together with the random morphisms

$$p(x) \in \mathrm{Hom}(\{e\}, X), \quad p(x'|x) \in \mathrm{Hom}(X, X)$$

so that: a) $p(x)$ is invariant with respect to the composition by $p(x'|x)$ i.e.

(4.28) $$p(x'|x) \circ p(x) = p(x') \qquad \text{(stationarity)}$$

b) we have

$$\lim_{n\to\infty} \frac{1}{n} \sum_{k=0}^{n-1} p(x'|x_k) \circ p(x_k|x_{k-1}) \circ \ldots \circ p(x_1|x) = p(x') . \quad (4.29)$$

(ergodicity)

If the entropy of the morphism $p(x'|x)$ by means of the standard morphism $p(x)$ i.e.

$$H = - \sum_{\substack{x \in X \\ x' \in X}} p(x) p(x'|x) \log p(x'|x)$$

satisfies the inequality

$$H < \mathbf{C} \quad (4.30)$$

where

$$\mathbf{C} = \sup_{\tilde{p}(x)} \left\{ \sum_{\substack{x \in X \\ y \in Y}} \tilde{p}(x) p(y|x) \log \frac{p(y|x)}{\sum_{x \in X} \tilde{p}(x) p(y|x)} \right\} \quad (4.31)$$

(the supremum being taken over all possible standard morphisms with the ending X) then for every $\varepsilon > 0$ there is the natural number $n_0(\varepsilon)$ such that for every $n \geqslant n_0(\varepsilon)$ there is a set

$$\tilde{A}_n \subset X^n = \prod_{k=1}^{n} X_{(k)} \qquad (X_{(k)} = X)$$

and a partition $\pi(Y^n)$ of the set

$$Y^n = \prod_{k=1}^{n} Y_{(k)} \qquad (Y_{(k)} = Y)$$

for which the restriction to the set $\tilde{A}_n$ of the extension to the partition $\pi(Y^n)$ of the product morphism of order n generated by $p(y|x)$ is ε-deterministic. At the same time, the set $\tilde{A}_n$ coincides ε-everywhere with the set X^n namely

$$p_n(X^n - \tilde{A}_n) < \varepsilon \tag{4.32}$$

where we denoted by p_n the extension to the partition

$$\{\tilde{A}_n, X^n - \tilde{A}_n\}$$

of the Markov morphism of order $n-1$ generated by the morphisms $p(x)$ and $p(x'|x)$;

PROOF: Let μ be the stationary ergodic Markov measure generated by the random morphisms $p(x)$ and $p(x'|x)$. The McMillan classical theorem implies that the sequence

$$\left(-\frac{1}{n}\log\mu(X_n)\right)_{1\leqslant n<\infty}$$

converges in probability towards the entropy H corresponding to the measure μ. Therefore for n sufficiently large we have

$$\mu\left(\frac{1}{n}\log\mu(X_n) + H \geqslant -\varepsilon\right) \geqslant 1-\varepsilon .$$

Let $\tilde{X}_n^{(1)},\dots,\tilde{X}_n^{(N')}$ be those cylinder sets of the form

$$X_n = [x_1,\dots,x_n]$$

satisfying the inequality

$$\frac{1}{n}\log\mu(X_n) + H \geq -\varepsilon\,.$$

We have

$$\sum_{i=1}^{N'}\mu(\tilde{X}_n^{(i)}) \geq 1-\varepsilon \tag{4.33}$$

$$\mu(\tilde{X}_n^{(i)}) \geq 2^{-n(H+\varepsilon)}\,, \quad (i = 1,\ldots,N')\,. \tag{4.34}$$

From (4.34) we have

$$1 \geq \sum_{i=1}^{N'}\mu(\tilde{X}_n^{(i)}) \geq N' 2^{-n(H+\varepsilon)}$$

i.e.

$$N' \leq 2^{n(H+\varepsilon)}\,. \tag{4.35}$$

On the other hand from the theorem 4.1 it follows that for $n \geq n_0(\varepsilon)$ there exist N cylinder sets $X_n^{(1)},\ldots,X_n^{(N)}$ and N disjoint sets $Y^{(1)},\ldots,Y^{(N)}$ every set $Y^{(i)}$ being a union of cylinder sets of the form

$$Y_n = [y_1,\ldots,y_n]$$

so that

$$\nu_{X_n^{(i)}}(Y^{(i)}) > 1-\varepsilon\,, \quad (i = 1,\ldots,N) \tag{4.36}$$

and

$$N > 2^{n(C-\varepsilon)} . \tag{4.37}$$

Let us suppose without any loss of generality that

$$C - H > 2\varepsilon .$$

From (4.35) and (4.37) it results

$$N' \leqslant 2^{n(H+\varepsilon)} < 2^{n(C-\varepsilon)} < N \tag{4.38}$$

be denoted by $\tilde{\bar{x}}_n^{(i)}$ the projection of $\tilde{X}_n^{(i)}$ on the set X^n i.e.

$$\tilde{\bar{x}}_n^{(i)} = (x_1^{(i)}, \dots, x_n^{(i)})$$

if

$$\tilde{X}_n^{(i)} = [x_1^{(i)}, \dots, x_n^{(i)}] , \quad (i = 1, \dots, N')$$

and by $\bar{x}_n^{(j)}$ the projection of $X_n^{(j)}$ on the set X^n i.e.

$$\bar{x}_n^{(j)} = (x_1^{(j)}, \dots, x_n^{(j)})$$

if

$$X_n^{(j)} = [x_1^{(j)}, \dots, x_1^{(j)}] , \quad (j = 1, \dots, N) .$$

Let us put

$$\tilde{A}_n = \{\tilde{\bar{x}}_n^{(1)}, \dots, \tilde{\bar{x}}_n^{(N')}\}$$

$$\bar{A}_n = \{\bar{x}_n^{(1)},\dots,\bar{x}_n^{(N')}\} \subset \{\bar{x}_n^{(1)},\dots,\bar{x}_n^{(N)}\} = A_n$$

where the last inclusion is possible owing to the inequality (4.38)

Given the deterministic morphism

$$\bar{p}(\bar{x}_n|\tilde{x}_n) : \tilde{A}_n \longrightarrow \bar{A}_n$$

so that

$$\bar{p}(\bar{x}_n^{(i)}|\tilde{x}_n^{(i)}) = 1 \qquad (i = 1,\dots,N'). \tag{4.39}$$

According to the theorem 4.1 there is a ε-deterministic morphism associated with the morphism $p(y|x)$,

$$p(\bar{y}_n|\bar{x}_n) \in \mathrm{Hom}(A_n, \pi(Y^n))$$

where

$$\pi(Y^n) = \{\bar{y}^{(1)},\dots,\bar{y}^{(N+1)}\}$$

and let there be the restriction of this ε-deterministic morphism to the set $\bar{A}_n$ i.e. let the morphism

$$\bar{p}(\bar{y}_n|\bar{x}_n) \in \mathrm{Hom}(\bar{A}_n, \pi(Y^n))$$

be the restriction of this ε-deterministic morphism to the set $\bar{A}_n$.

Let us now define the morphism

$$\tilde{p}(\bar{y}_n|\tilde{x}_n) = \bar{p}(\bar{y}_n|\bar{x}_n)\circ p(\bar{x}_n|\tilde{x}_n) .$$

This is an ε-deterministic morphism and

$$\tilde{p}(\bar{y}_n|\tilde{x}_n) \in \mathrm{Hom}(\tilde{A}_n, \pi(Y^n)) .$$

Indeed, for every

$$\tilde{x}_n^{(i)} \in \tilde{A}_n , \quad (i = 1,\ldots,N')$$

we have

$$p(\bar{x}_n^{(i)}|\tilde{x}_n^{(i)}) = 1$$

and according to the last theorem proved above

$$\bar{y}^{(i)} \in \pi(Y^n)$$

exists

so that

$$\bar{p}(\bar{y}^{(i)}|\bar{x}_n^{(i)}) = p(\bar{y}^{(i)}|\bar{x}_n^{(i)}) > 1 - \varepsilon$$

therefore

$$\tilde{p}(\bar{y}^{(i)}|\tilde{x}_n^{(i)}) = \sum_{k=1}^{N'} \bar{p}(\bar{y}^{(i)}|\bar{x}_n^{(k)})\bar{p}(\bar{x}_n^{(k)}|\tilde{x}_n^{(i)}) =$$

$$= \bar{p}(\bar{y}^{(i)}|\bar{x}_n^{(i)})\bar{p}(\bar{x}_n^{(i)}|\tilde{x}_n^{(i)}) = \bar{p}(\bar{y}^{(i)}|\bar{x}_n^{(i)}) > 1 - \varepsilon .$$

On the other hand, given the Markov morphism of

order $(n-1)$

$$p_n(x_1,\ldots,x_n)\in \mathrm{Hom}(\{e\},X^n)$$

generated by $p(x)$ and $p(x'|x)$ or the restriction to X^n of the Markov measure μ generated by $p(x)$ and $p(x'|x)$. Let us consider the extension to the partition

$$\{\tilde{A}_n, X^n - \tilde{A}_n\}$$

of the Markov morphism $p_n(x_1,\ldots,x_n)$ i.e. the morphism

$$p_n:\{e\}\longrightarrow\{\tilde{A}_n, X^n - \tilde{A}_n\}.$$

We have seen that the cylinder sets $\tilde{X}_n^{(i)}, (i=1,\ldots,N')$ satisfy the inequality (4.4). It follows therefore that for other

$$(\mathrm{card}\, X)^n - N'$$

possible cylinder sets

$$\tilde{X}_n^{(j)}, \quad (j = N'+1,\ldots,(\mathrm{card}\, X)^n)$$

we have

$$\mu\left(\bigcup_{j=N'+1}^{(\mathrm{card}\, X)^n} \tilde{X}_n^{(j)}\right) < \varepsilon$$

i.e.

$$p_n(X^n - \tilde{A}_n) < \varepsilon \qquad \text{q.e. d.}$$

The deterministic morphism (4.39) will be called <u>n -dimensional code</u> in the set X. It realizes a <u>codification</u> in the product set X^0.

Let us notice that in the theorems 4.1 and 4.2 the random morphism $p(y|x)$ was chosen arbirtarily. X in particular may itself be taken to be a product set, i.e. $X = Z^n$ in which case the theorems 4.1 and 4.2 may be formulated without any difficulty, thus demonstrating the necessary conditions for the reduction of the random morphism

$$p(y|z_1, z_2, \dots, z_m) \in \mathrm{Hom}(Z^m, Y)$$

to an ε-deterministic morphism.

Let us demonstrate some applications of the theory presented above. To start with, let us consider a stationary discrete memoryless noisy communication system for which the input signals form a stationary ergodic Markov chain determined by the morphisms

$$p(x) \in \mathrm{Hom}(\{e\}, X), \quad p(x'|x) \in \mathrm{Hom}(X, X)$$

such that the relations (4.28) and (4.29) hold. Then if the entropy of this Markov chain

$$H = - \sum_{\substack{x \in X \\ x' \in X}} p(x) p(x'|x) \log p(x'|x)$$

satisfies the inequality (4.30) then according to the theorem 4.2 for every $\varepsilon > 0$ a natural number $n_0(\varepsilon)$ exists so that for every $n \geq n_0(\varepsilon)$ there is an n-dimensional code in the set X. As a consequence of this codification, the identification of the output with the input, may be made with an error smaller than ε.

The number $\mathbf{C}$ depends only on the perturbation $p(y|x)$ of the communication channel and is called the capacity of the respective channel.

If we replace the morphism $p(y|x)$ by the morphism having the form $p(y|x_1,\dots,x_m)$ then the theorem 4.2 gives us the codification theorem for a channel with the memory $\mathbf{m}$.

Although from the theorem 4.2 we obtain the following codification theorem for a stationary random automata:

THEOREM 4.3: Let there be an homogeneous random automata and let us suppose that

$$\lim_{n\to\infty}\frac{1}{n}\sum_{k=0}^{n-1}p_{(k)}(a',x'|a,x) = p_0(a',x')$$

where $p_{(k)}(a',x'|a,x)$ represents the k-th power of the morphism of transition of this automaton. If

$$-\sum_{a,a',x,x'}p_0(a,x)p(a',x'|a,x)\log p(a',x'|a,x) < \mathbf{C} \qquad (4.40)$$

where

$$\mathbf{C} = \sup_{p(a,x)}\left\{\sum_{a,x,y}p(a,x)p(y|a,x)\log\frac{p(y|a,x)}{\sum_{a,x}p(a,x)p(y|a,x)}\right\}$$

(the supremum being taken over all possible standard morphisms with the ending $A\times X$) then for every $\varepsilon>0$ there exists a natural number $n_0(\varepsilon)$ so that for every $n>n_0(\varepsilon)$ there is an n-dimensional code so that as a consequence of this codification, the identification of the output of the automaton with the pairs state-input of the same automaton may be made with an error smaller than ε.

It should be noted that (4.40) expresses a connection between the transition morphism and the output morphism of the random automaton namely the connection between the perturbation on the transition channel and the perturbation on the output channel of the random automaton.

Throughout the chapter it will be assumed that an FR-category $\mathcal{C}$ of $\sum$-type well-equipped is given. All the objects and random morphisms dealing with it are supposed to belong to this category.

Let

$$(4.41) \qquad X^n \xrightarrow{p(x_1',\ldots,x_n'|x_1,\ldots,x_n)} X^n$$

be a random morphism and let us define the distance between two arbitrary elements

$$(x_1,\ldots,x_n)\in X^n \ , \quad (x_1',\ldots,x_n')\in X^n$$

by the equality

$$(4.42) \qquad d(x_1',\ldots,x_n';x_1,\ldots,x_n) = \mathrm{card}\{j|x_j' \neq x_j \ , \ j = 1,\ldots,n\} \ .$$

Thus the distance between two arbitrary elements belonging to X^n is just the number of their distinct components.

Let us also define the number

$$e = \max_{(x'_1,\ldots,x'_n)\in X^n} \max_{(x_1,\ldots,x_n)\in X^n} \cdot$$

$$\cdot \operatorname{card}\{j \mid x'_j \neq x_j; j = 1,\ldots,n; p(x'_1,\ldots,x'_n \mid x_1,\ldots,x_n) > 0\} . \tag{4.43}$$

The non-negative number e may be interpreted as the maximum number of alter components with respect to the random morphism

$$p(x'_1,\ldots,x'_n \mid x_1,\ldots,x_n)$$

given above. Of course, for an arbitrary identical deterministic morphism the corresponding number e is always equal to zero.

Let us choose an arbitrary element $(x^1_1,\ldots,x^1_n)\in X^n$ and pick up an element $(x^2_1,\ldots,x^2_n)\in X^n$ such that

$$d(x^2_1,\ldots,x^2_n \mid x^1_1,\ldots,x^1_n) > 2e$$

supposing that $e < n$.

Furthermore let an element $(x^3_1,\ldots,x^3_n)\in X^n$ be selected such that

$$d(x^3_1,\ldots,x^3_n ; x^1_1,\ldots,x^1_n) > 2e$$

$$d(x^3_1,\ldots,x^3_n ; x^2_1,\ldots,x^2_n) > 2e .$$

Let us choose the element $(x^j_1,\ldots,x^j_n)\in X^n$ step by step, so that

$$d(x_1^j,\ldots,x_n^j;x_1^i,\ldots,x_n^i) > 2e \qquad (i = 1,\ldots,j-1).$$

The process may be continued for only a finite number of steps and let

(4.44) $$\tilde{X}^n = \{x_1^j,\ldots,x_n^j \mid j = 1,\ldots,N\}$$

be the selected set. According to the construction device given above we have

(4.45) $$d = \min_{\substack{i,j=1,\ldots,N \\ i \neq j}} d(x_1^j,\ldots,x_n^j;x_1^i,\ldots,x_n^i) > 2e .$$

Denoting by

$$S(x_1^j,\ldots,x_n^j;e)$$

the sphere having the centre $(x_1^j,\ldots,x_n^j)$ and the radius e i.e

(4.46) $$S(x_1^j,\ldots,x_n^j;e) = \{(x_1,\ldots,x_n) \mid d(x_1,\ldots,x_n;x_1^j,\ldots,x_n^j) \leq e\}.$$

The correspondence

(4.47) $$\tilde{X}^n \xrightarrow{\;p(S(x_1^k,\ldots,x_n^k;e)\mid x_1^i,\ldots,x_n^i)\;} \Psi(X^n)$$

is a deterministic morphism, i.e.

(4.48) $$p(S(x_1^k,\ldots,x_n^k;e)\mid x_1^i,\ldots,x_n^i) = \delta_{ik} = \begin{cases} 1 & \text{if} \quad i = k \\ 0 & \text{if} \quad i \neq k \end{cases}$$

where $\Psi(X^n)$ is the set of subsets of X^n defined by the spheres corresponding to the set $\tilde{X}^n$ i.e.

$$\Psi(X^n) = \{S(x_1^j,\dots,x_n^j;e) | j = 1,\dots,N\} . \qquad (4.49)$$

In the considerations made above the number e of the maximum positions that may be altered with respect to the random morphism (4.41) in an arbitrary succession of length n from X^n was a priori given. Afterwards the minimum distance d in the set $\tilde{X}^n$ was chosen so that $d > 2e$. Let us start with an initial minimum distance d and let us select a subset

$$\tilde{X}^n \subset X^n$$

such that the distance between two arbitrary elements of the set should be at least d. Then letting by

$$e^* = \left[\frac{d-1}{2}\right]$$

let us consider the set of the spheres centred in the elements of the set $\tilde{X}^n$ having the rays equal to e^*, i.e.

$$\Psi(X^n) = \{S(x_1^j,\dots,x_n^j;e^*) | j = 1,\dots,N\} .$$

Let us put

$$\varepsilon = \max_{1 \le j \le N} p(X^n - S(x_1^j,\dots,x_n^j;e^*) | x_1^j,\dots,x_n^j) . \qquad (4.51)$$

Then the correspondence

$$\tilde{X}^n \xrightarrow{p(S(x_1^j,\dots,x_n^j;e^*) | x_1^j,\dots,x_n^j)} \Psi(X^n)$$

is an ε-deterministic morphism, i.e.

$$p(S(x_1^j,\ldots,x_n^j;e^*)|x_1^j,\ldots,x_n^j) > 1-\varepsilon \tag{4.52}$$

whatever be $j=1,\ldots,N$.

It is of prime importance to have a rule which allows us to realize an effective selection of the set $\tilde{X}^n$ so that the number of the elements of this set should be sufficiently large.

Let us suppose that the set X consists of two elements denoted by 0 and 1 respectively. This first method for the construction of the subsets $\tilde{X}^n \subset X^n$ is based on the remarkable properties of Hadamard matrices. This method is generally applied in almost all interesting situations. If the Hadamard matrix $\mathcal{H}_n$ is given there exists a subset $A_n \subset X^n$ having $2n$ elements so that the minimum distance in this set is $n/2$. The set A_n may be constructed in the following manner: let us form the $2n$ vectors with n components $v_1,\ldots,v_n,-v_1,\ldots,-v_n$ where $v_1,\ldots,v_n$ are the orthogonal rows of the matrix $\mathcal{H}_n$. Then in each of these we change the components 1 to 0 and -1 to 1. Since corresponding components of v_i and $-v_i$ are different the distance between the elements of X^n obtained from v_i and the element of X^n obtained from $-v_i$ is n. But since $\pm v_i$ and $\pm v_j$ are orthogonal if $i \neq j$ that they must match in one half of the components and differ in the other half. Therefore the corresponding elements from X^n are at the distance $n/2$.

We are interested in the possibility of constructing Hadamard's matrices (therefore the desired subset A_n) for the values of n sufficiently large. This possibility is a result of the simple observation that if $\mathcal{H}_n$ is a Hadamard matrix then

$$\mathcal{H}_{2n} = \begin{bmatrix} \mathcal{H}_n & \mathcal{H}_n \\ \mathcal{H}_n & -\mathcal{H}_n \end{bmatrix} \tag{4.53}$$

is a Hadamard too. It should also be noted that

$$\mathcal{H}_2 = \begin{bmatrix} 1 & 1 \\ 1 & -1 \end{bmatrix} \tag{4.54}$$

is a Hadamard matrix.

In the effective applications of the theory presented above by using the Hadamard matrices we select first a large set having the minimum distance $n/2$ and afterwards we pick up the desired set

$$\tilde{X}^n \subset A_n \subset X^n \tag{4.55}$$

such that the "error" ε given by (4.51) is minimum.

The selection of the set A_n can be made also by using the Bose-Chaudhuri-Hocquenghem codes or the modular representation of the linear codes (especially the McDonald codes). (For all these notions see W. Peterson's book: "Error Correcting codes", Wiley, New York, 1961.). The algebraic structure on the set X permits us this selection of first importance from the point of view of approximation of the random morphism by ε-deterministic morphisms.

Therefore we are able to construct a set $A_n \subset X^n$ having the desired minimum distance and we have chosen the set $\tilde{X}^n$ so that $\tilde{X}^n \subset A_n$. The corresponding random morphism (4.48) or (4.52) is a deterministic (respectively ε-deterministic) morphism. One problem arises very naturally, given an arbitrary element

$$(x_1, \ldots, x_n) \in X^n$$

How it is possible to recognize the sphere

$$S(x_1^+, \ldots, x_n^+; e) \tag{4.56}$$

quickly for which

$$(x_1, \ldots, x_n) \in S(x_1^+, \ldots, x_n^+; e) \tag{4.57}$$

and therefore the element $(x_1^+, \ldots, x_n^+) \in X^n$ corresponding with the probability 1 to $(x_1, \ldots, x_n)$ by means of the deterministic morphism (4.48) (or corresponding with the probability greater than $1-\varepsilon$ to $(x_1, \ldots, x_n)$ by means of the ε-deterministic morphism (4.52). We shall apply here the most rational algorithm of recognition. But first of all let us speak a little about the general formulation of this algorithm.

Let

$$E_n = \{s_1, \ldots, s_n\}$$

be a finite set having $n \geq 2$ distinguishable elements called entities and suppose that we want to recognize an unknown entity s of the set E_n. The set E_n itself is supposed to be known to us. Let us further suppose that it is not possible to observe the entity s directly but we may choose to observe the values taken on by some characteristics defining the respective entities. We want to recognize s by a limited reasonable number of observation.

Landa proposed such a strategy of recognition which was formalized in the paper [12]. According to this algorithm it is necessary at every moment to choose and to observe first of all such a characteristic supplying the largest amount of information i.e. eliminating the largest degree of uncertainty.

We intend to apply the considerations made above to the effective coding and decoding procedure. This is both a conciliation of the probabilistic and algebraic theory of codes and an approach towards a simple decoding device, using for this aim the most rational algorithm of recognition.

Let us consider an information source transmitting signals belonging to the alphabet A. Let us suppose that in A^m, we have N elements $s_1, \ldots, s_N$ i.e. N blocks of length m with the probabilities

$$p(s_i) \geq 0 , \quad (i = 1, \ldots, N) \tag{4.58}$$

such that

(4.59) $$\sum_{i=1}^{N} p(s_i) > 1 - \varepsilon'$$

or equivalently

$$\sum_{s \in A^m - \{s_1,\dots,s_N\}} p(s) < \varepsilon' .$$

Here $p(s_i)$ represents the transmission probability of the sequence of length m, $s_i \in A^m$.

Let there be a communication channel having as input and respectively output alphabet the same set X and the perturbation defined by the family of random morphisms

$$\mathrm{Hom}(X^n, X^n) = \{p(x'_1,\dots,x'_n | x_1,\dots,x_n)\} , \quad (n = 1,2,\dots) .$$

a) If $d = 2e+1$ is given let us choose the natural number n such that in the set X^n it should be possible to select (using eventually the Hadamard's matrices method or other method) a subset $A_n \subset X^n$ for which:

1°. the minimum distance is at least d

2°. $\mathrm{card}\, A_n > N$

Let us define the subset

$$\tilde{X}^n \subset A_n$$

having N elements chosen such that

(4.60) $$p_n(S(x_1,\dots,x_n;e)|x_1,\dots,x_n) \leqslant$$

$$\leq \min_{1\leq j\leq N} p_n(S(x_1^j,\dots,x_n^j;e)|x_1^j,\dots,x_n^j) \tag{4.60}$$

for every

$$(x_1,\dots,x_n)\in A_n - \tilde{X}^n .$$

Let us put

$$\varepsilon'' = \max_{1\leq j\leq N} p(X^n - S(x_1^j,\dots,x_n^j;e)|x_1^j,\dots,x_n^j) . \tag{4.61}$$

It is now possible to define the following code, i.e. the following deterministic morphism

$$\varphi = \varphi(N,m,d,n):A^m \longrightarrow X^n$$

so that

$$\varphi(s_j) = (x_1^j,\dots,x_n^j)\in\tilde{X}^n \tag{4.62}$$

for every $j = 1,\dots,N$ and

$$\varphi(s)\in X^n - \tilde{X}^n \tag{4.63}$$

for every

$$s\in A^m - \{s_1,\dots,s_N\} .$$

Let there be

$$\varepsilon = \varepsilon' + \varepsilon'' . \tag{4.64}$$

Thus receiving the output sequence $(x_1,..,x_n)\in X^n$ with an error smaller than ε'' it is possible to determine the sphere

$$S(x_1^i,\dots,x_n^i;e)$$

to which $(x_1,\dots,x_n)$ belongs and further with an error smaller than ε' it is possible to determine the input sequence that was transmitted.

Therefore with a total error smaller than ε it is possible to identify the input when the output was received.

b) If the natural number n is given, i.e. when the random morphism

$$p(x_1',\dots,x_n'|x_1,\dots,x_n)$$

is given, let us determine the number e so that

$$e = \max_{(x_1',\dots,x_n')\in X^n}\ \max_{(x_1,\dots,x_n)\in X^n} \operatorname{card}\cdot \tag{4.65}$$

$$\cdot\{j|x_j' \neq x_j, j = 1,\dots,n; p(x_1',\dots,x_n'|x_1,\dots,x_n) > \varepsilon\}$$

and let us take $d = 2e+1$.

Further let us pick up the set $A_n \subset X^n$ having the minimum distance of at least d. If

$$N < \operatorname{card} A_n \tag{4.66}$$

let us define the desired subset $\tilde{X}^n$ and the code φ as at the point a).

Nevertheless our aim is to have a simple decoding procedure. The most rational algorithm of recognition will be applied for the speedy identification of the sphere to which the received output sequence $(x_1,..,x_n)$ belongs but the nature and order of the components of the received output sequence must be verified to recognize the corresponding sphere i.e. to recognize the corresponding input sequence. In this way all the difficulties are shifted to the coding where the most rational algorithm of recognition must be elaborate. This algorithm of recognition (i.e. its final graph of recognition) may be utilized at the output of the communication channel even by less highly qualified personnel. To utilize the most rational algorithm of recognition for the decoding, no calculation is involved.

The following example will serve to clarify the above considerations, e.g. 13 input sequences $s_i,(1\leqslant i\leqslant 13)$ having the probabilities

$$p(s_1)=0.049996;\ p(s_2)=0.049996;\ p(s_3)=0.099992;\ p(s_4)=0.099992;$$
$$p(s_5)=0.299976;\ p(s_6)=0.049996;\ p(s_7)=0.199984;\ p(s_8)=0.149988;$$
$$p(s_9)=0.000040;\ p(s_{10})=0.000015;\ p(s_{11})=0.000010;\ p(s_{12})=0.000010;$$
$$p(s_{13})=0.000005. \tag{4.67}$$

Let us select the most probable subset composed by the first eight sequences. As a consequence of this selection the error made will be equal to 0.00008 . Therefore $\varepsilon'=0.0008, N=8$

Let us consider a binary communication channel having as input and output alphabet the same binary set $X=\{0,1\}$. Let us take $n=8$. We select, using Hadamard matrices techniques a set

$$A_8 \subset X^8$$

having the minimum distance $d = 4$. Thus according to the considera-tion made above we consider the Hadamard matrix

$$\mathcal{H}_8 = \begin{bmatrix} 1 & 1 & 1 & 1 & 1 & 1 & 1 & 1 \\ 1 & -1 & 1 & -1 & 1 & -1 & 1 & -1 \\ 1 & 1 & -1 & -1 & 1 & 1 & -1 & -1 \\ 1 & -1 & -1 & 1 & 1 & -1 & -1 & 1 \\ 1 & 1 & 1 & 1 & -1 & -1 & -1 & -1 \\ 1 & -1 & 1 & -1 & -1 & 1 & -1 & 1 \\ 1 & 1 & -1 & -1 & -1 & -1 & 1 & 1 \\ 1 & -1 & -1 & 1 & -1 & 1 & 1 & -1 \end{bmatrix}.$$

We construct the corresponding set A_8 having as elements

$$\begin{array}{cc} 0\,0\,0\,0\,0\,0\,0\,0 & 1\,1\,1\,1\,1\,1\,1\,1 \\ 0\,1\,0\,1\,0\,1\,0\,1 & 1\,0\,1\,0\,1\,0\,1\,0 \\ 0\,0\,1\,1\,0\,0\,1\,1 & 1\,1\,0\,0\,1\,1\,0\,0 \\ 0\,1\,1\,0\,0\,1\,1\,0 & 1\,0\,0\,1\,1\,0\,0\,1 \\ 0\,0\,0\,0\,1\,1\,1\,1 & 1\,1\,1\,1\,0\,0\,0\,0 \\ 0\,1\,0\,1\,1\,0\,1\,0 & 1\,0\,1\,0\,0\,1\,0\,1 \\ 0\,0\,1\,1\,1\,1\,0\,0 & 1\,1\,0\,0\,0\,0\,1\,1 \\ 0\,1\,1\,0\,1\,0\,0\,1 & 1\,0\,0\,1\,0\,1\,1\,0 \end{array} \tag{4.68}$$

with the minimum distance $d = 8/2 = 4$.

Let us suppose that on the given communication channel the perturbation is such that the simple substitution error (i.e. the perturbation of a single arbitrary signal in a given succession of length 8) has a probability almost equal to 1. Let us suppose that for the sequence mentioned above in (4.68) we have

$$
\begin{aligned}
p(S(00000000;1)|00000000) &= 0.99995\\
p(S(01010101;1)|01010101) &= 0.99997\\
p(S(00110011;1)|00110011) &= 0.99890\\
p(S(01100110;1)|01100110) &= 0.99996\\
p(S(00001111;1)|00001111) &= 0.99973\\
p(S(01011010;1)|01011010) &= 0.99996\\
p(S(00111100;1)|00111100) &= 0.99890\\
p(S(01101001;1)|01101001) &= 0.99997\\
p(S(11111111;1)|11111111) &= 0.99831\\
p(S(10101010;1)|10101010) &= 0.98902\\
p(S(11001100;1)|11001100) &= 0.99998\\
p(S(10011001;1)|10011001) &= 0.99978\\
p(S(11110000;1)|11110000) &= 0.99994\\
p(S(10100101;1)|10100101) &= 0.99886\\
p(S(11000011;1)|11000011) &= 0.99996\\
p(S(10010110;1)|10010110) &= 0.99799\ .
\end{aligned}
\qquad (4.69)
$$

According to these values we pick up the subset

$$\tilde{X}^8 \subset A_8$$

composed of the following successions selected in decreasing order of the probabilities given above

$$\tilde{X}^8 = \{11001100, 01010101, 01101001, 01100110, 01011010, 11000011, 00000000, 11110000\} . \quad (4.70)$$

Of course, in our case

$$\varepsilon'' = 0.00006 , \quad e = 1 .$$

Let us suppose that the effect of the perturbation (i.e. the transition probabilities) on the successions belonging to $\tilde{X}^8$ is

$$\begin{aligned}
&p(11001100|11001100) = 0.7999840 \\
&p(01001100|11001100) = 0.0499990 \\
&p(10001100|11001100) = 0.0399992 \\
&p(11011100|11001100) = 0.0000000 \\
&p(11101100|11001100) = 0.0099998 \\
&p(11000100|11001100) = 0.0199996 \\
&p(11001000|11001100) = 0.0199996 \\
&p(11001101|11001100) = 0.0199996 \\
&p(11001110|11001100) = 0.0399992
\end{aligned} \quad (4.71)$$

$$\begin{aligned}
p(01010101|01010101) &= 0.4999850\\
p(00010101|01010101) &= 0.0499985\\
p(10010101|01010101) &= 0.0199994\\
p(01000101|01010101) &= 0.1999940\\
p(01110101|01010101) &= 0.0299991\\
p(01010001|01010101) &= 0.0099997\\
p(01011101|01010101) &= 0.0399988\\
p(01010101|01010101) &= 0.0999970\\
p(01010111|01010101) &= 0.0499985
\end{aligned} \tag{4.72}$$

$$\begin{aligned}
p(01101001|01101001) &= 0.7999760\\
p(00101001|01101001) &= 0.0199994\\
p(11101001|01101001) &= 0.0099997\\
p(01111001|01101001) &= 0.0099997\\
p(01001001|01101001) &= 0.0099997\\
p(01101101|01101001) &= 0.0499985\\
p(01100001|01101001) &= 0.0099997\\
p(01101000|01101001) &= 0.0499985\\
p(01101011|01101001) &= 0.0399988
\end{aligned} \tag{4.73}$$

$$\begin{aligned}
p(01100110|01100110) &= 0.7499700\\
p(00100110|01100110) &= 0.0000000\\
p(11100110|01100110) &= 0.0499980\\
p(01110110|01100110) &= 0.0499980\\
p(01000110|01100110) &= 0.0299988\\
p(01100010|01100110) &= 0.0199992\\
p(01101110|01100110) &= 0.0499980\\
p(01101111|01100110) &= 0.0499980\\
p(01100100|01100110) &= 0.0000000
\end{aligned} \tag{4.74}$$

$$p(01011010|01011010) = 0.8999640$$
$$p(00011010|01011010) = 0.0099996$$
$$p(11011010|01011010) = 0.0199992$$
(4.75)
$$p(01001010|01011010) = 0.0099996$$
$$p(01111010|01011010) = 0.0199992$$
$$p(01011110|01011010) = 0.0000000$$
$$p(01010010|01011010) = 0.0199992$$
$$p(01011011|01011010) = 0.0199992$$
$$p(01011000|01011010) = 0.0000000$$

$$p(11000011|11000011) = 0.8499660$$
$$p(10000011|11000011) = 0.0000000$$
$$p(01000011|11000011) = 0.0299988$$
(4.76)
$$p(11010011|11000011) = 0.0399984$$
$$p(11100011|11000011) = 0.0000000$$
$$p(11000111|11000011) = 0.0099996$$
$$p(11001011|11000011) = 0.0199992$$
$$p(11000001|11000011) = 0.0299988$$
$$p(11000010|11000011) = 0.0199992$$

$$p(00000000|00000000) = 0.5999700$$
$$p(01000000|00000000) = 0.0999950$$
$$p(10000000|00000000) = 0.0499975$$
(4.77)
$$p(00010000|00000000) = 0.0099995$$
$$p(00100000|00000000) = 0.0099995$$
$$p(00000100|00000000) = 0.0199990$$
$$p(00001000|00000000) = 0.0799960$$
$$p(00000001|00000000) = 0.0299985$$
$$p(00000010|00000000) = 0.0099995$$

$$p(11110000|11110000) = 0.6499610$$
$$p(10110000|11110000) = 0.0499970$$
$$p(01110000|11110000) = 0.0999940$$
$$p(11100000|11110000) = 0.0499970$$
$$p(11010000|11110000) = 0.0199988 \qquad (4.78)$$
$$p(11110100|11110000) = 0.0199988$$
$$p(11111000|11110000) = 0.0199988$$
$$p(11110001|11110000) = 0.0199988$$
$$p(11110010|11110000) = 0.0399976 .$$

In this way we have written explicitly the components of

$$p(S(x_1x_2x_3x_4x_5x_6x_7x_8;1)|x_1x_2x_3x_4x_5x_6x_7x_8)$$

for every

$$x_1x_2x_3x_4x_5x_6x_7x_8 \in \tilde{X}^8 .$$

According to these values it is necessary to consider the following code

$$\varphi(s_1) = 00000000 ; \quad \varphi(s_2) = 01010101 ; \quad \varphi(s_3) = 01100110 ;$$
$$\varphi(s_4) = 01101001 ; \quad \varphi(s_5) = 01011010 ; \quad \varphi(s_6) = 11110000 ; \qquad (4.79)$$

$$\varphi(s_7) = 11000011\,; \quad \varphi(s_8) = 11001100\,.$$

Let us suppose that with an error of $\varepsilon' = 8.10^{-5}$ only

$$s_1, s_2, s_3, s_4, s_5, s_6, s_7, s_8$$

are transmitted. Thus their probabilities will be

$$p(s_1) = 0.05\,; \quad p(s_2) = 0.05\,; \quad p(s_3) = 0.10\,; \quad p(s_4) = 0.10\,;$$
(4.80)
$$p(s_5) = 0.30\,; \quad p(s_6) = 0.05\,; \quad p(s_7) = 0.20\,; \quad p(s_8) = 0.15\,.$$

We can suppose also with an error $\varepsilon' = 6.10^{-5}$ that if we transmitted an arbitrary succession

$$x_1 x_2 x_3 x_4 x_5 x_6 x_7 x_8 \in \tilde{X}^8$$

we will receive at the channel output on the communication, only elements belonging to

$$S(x_1 x_2 x_3 x_4 x_5 x_6 x_7 x_8; 1)\,.$$

Then according to the values (4.71) - (4.78) we have

$p(00000000|00000000) = 0.60$
$p(01000000|00000000) = 0.10$
$p(10000000|00000000) = 0.05$
$p(00010000|00000000) = 0.01$
$p(00100000|00000000) = 0.10$
$p(00000100|00000000) = 0.02$
$p(00001000|00000000) = 0.08$
$p(00000001|00000000) = 0.03$
$p(00000010|00000000) = 0.01$

$p(01010101|01010101) = 0.50$
$p(00010101|01010101) = 0.05$
$p(11010101|01010101) = 0.02$
$p(01000101|01010101) = 0.20$
$p(01110101|01010101) = 0.03$
$p(01010001|01010101) = 0.01$
$p(01011101|01010101) = 0.04$
$p(01010100|01010101) = 0.10$
$p(01010111|01010101) = 0.05$

$p(01100110|01100110) = 0.75$
$p(00100110|01100110) = 0.00$
$p(11100110|01100110) = 0.05$
$p(01110110|01100110) = 0.05$
$p(01000110|01100110) = 0.03$
$p(01100010|01100110) = 0.02$
$p(01101110|01100110) = 0.05$
$p(01100100|01100110) = 0.05$
$p(01100111|01100110) = 0.00$

$p(01101001|01101001) = 0.80$
$p(00101001|01101001) = 0.02$
$p(11101001|01101001) = 0.01$
$p(01111001|01101001) = 0.01$
$p(01001001|01101001) = 0.01$
$p(01101101|01101001) = 0.05$
$p(01100001|01101001) = 0.01$
$p(01101000|01101001) = 0.05$
$p(01101011|01101001) = 0.04$

$p(01011010|01011010) = 0.90$
$p(00011010|01011010) = 0.01$
$p(11011010|01011010) = 0.02$
$p(01001010|01011010) = 0.01$
$p(01111010|01011010) = 0.02$
$p(01011110|01011010) = 0.00$
$p(01010010|01011010) = 0.02$
$p(01011011|01011010) = 0.02$
$p(01011000|01011010) = 0.00$

$p(11110000|11110000) = 0.65$
$p(10110000|11110000) = 0.05$
$p(01110000|11110000) = 0.10$
$p(11100000|11110000) = 0.05$
$p(11010000|11110000) = 0.05$
$p(11110100|11110000) = 0.02$
$p(11111000|11110000) = 0.02$
$p(11110001|11110000) = 0.02$
$p(11110010|11110000) = 0.04$

$p(11000011|11000011) = 0.85$
$p(10000011|11000011) = 0.00$
$p(01000011|11000011) = 0.03$
$p(11010011|11000011) = 0.04$
$p(11100011|11000011) = 0.00$
$p(11000111|11000011) = 0.01$
$p(11001011|11000011) = 0.02$
$p(11000001|11000011) = 0.03$
$p(11000010|11000011) = 0.02$

$p(11001100|11001100) = 0.80$
$p(01001100|11001100) = 0.05$
$p(10001100|11001100) = 0.04$
$p(11011100|11001100) = 0.00$
$p(11101100|11001100) = 0.01$
$p(11000100|11001100) = 0.02$
$p(11001000|11001100) = 0.02$
$p(11001101|11001100) = 0.02$
$p(11001110|11001100) = 0.04$

We shall now construct the most rational algorithm of recognition under the condition given above. Again the entities will be the eight spheres with the radius 1, having as centres the elements of the set $\tilde{X}^8$. The initial (a priori) probabilities are the following: the initial probability of the entity $S(\varphi(s_i);1)$ is equal to the probability of the element s_i whichever be $1 \leqslant i \leqslant 8$ where φ is the code (4.79). An arbitrary sequence $x_1 x_2 x_3 x_4 x_5 x_6 x_7 x_8$ received at the output of the communication channel will belong (a priori) with the probability $p(s_i)$ given by (4.70) to the sphere $S(\varphi(s_i);1)$ for every $1 \leqslant i \leqslant 8$. Let us consider the first, second, third and fourth pairs of components in the arbitrary sequence of length 8. Let 1, 2, 3, 4 respectively denote these characteristics. Every characteristic takes on only four distinct values, namely

$$00, \quad 01, \quad 10, \quad 11.$$

Following the method of the algorithm of recognition we see that the probabilities of the different values of the first characteristic are as follows:

$$p_1(00) = 0.85p(s_1) + 0.05p(s_2) + 0.02p(s_4) + 0.01p(s_5) = 0.050$$

$$p_1(01) = 0.10p(s_1) + 0.93p(s_2) + 0.95p(s_3) + 0.97p(s_4) + 0.97p(s_5) + $$
$$+ 0.05p(s_6) + 0.05p(s_8) = 0.545$$

$$p_1(10) = 0.05p(s_1) + 0.10p(s_6) + 0.03p(s_7) + 0.04p(s_8) = 0.020$$

$$p_1(11) = 0.02p(s_2) + 0.05p(s_3) + 0.01p(s_4) + 0.02p(s_5) + 0.85p(s_6) + $$
$$+ 0.97p(s_7) + 0.91p(s_8) = 0.386$$

For the other characteristics we have

$$p_2(00) = 0.404\ ,\quad p_2(01) = 0.341\ ,\quad p_2(10) = 0.199\ ,\quad p_2(11) = 0.056$$

$$p_3(00) = 0.295\ ,\quad p_3(01) = 0.148\ ,\quad p_3(10) = 0.400\ ,\quad p_3(11) = 0.157$$

$$p_4(00) = 0.257\ ,\quad p_4(01) = 0.145\ ,\quad p_4(10) = 0.390\ ,\quad p_4(11) = 0.208$$

The corresponding entropies will be

$$H_1 = 1.3363\ ,\quad H_2 = 1.7540\ ,\quad H_3 = 1.8757\ ,\quad H_4 = 1.9088$$

Therefore we shall verify first of all the fourth characteristic. If we obtain the value 00 for it then we still have seven alternatives, namely

00000000 V 01000000 V 10000000 V 00010000 V 00100000 V

V 00000100 V 00001000 , 01010100 , 01100100 , 01101000 ,

01011000 , 11110000 V 01110000 V 10110000 V 11010000 V

V 11100000 V 11110100 V 11111000 V 11001100 V 01001100 V

V 10001100 V 11011100 V 11101100 V 11000100 V 11001000

corresponding respectively to $s_1, s_2, s_3, s_4, s_5, s_6, s_8$ and having, according to (4.70), the probabilities

0.063 , 0.063 , 0.125 , 0.125 , 0.375 , 0.063 , 0.186

respectively.

When estimating the information supplied by the other characteristics in this situation, we obtain the following results.

$$H_1 = 1.1629 , \quad H_2 = 1.7561 , \quad H_3 = 1.7609 .$$

Let us verify the third characteristic. If we obtain the value 00 we may have two alternatives, namely:

00000000 V 01000000 V 10000000 V 00010000 V 00100000 ,

11110000 V 01110000 V 10110000 V 11010000 V 11100000

corresponding respectively to s_1 and s_6 having, according to (4.70), the probabilities 0.500 . In this situation a new computation shows us the values taken on by the entropy for the first and second characteristics:

$$H_1 = 1.6098 , \quad H_2 = 1.5013 .$$

We thus verify the first characteristic and the

value 00 for it implies s_1.

In this manner we obtain one possible method of the most rational algorithm of recognition. Dealing similarly all other possible situations we finally obtain the following algorithm of recognition including all these possibilities (see the next page).

Using this algorithm, with an error $\varepsilon = 14.10^{-5}$ we are able to recognize the input correctly. The mean length of the final graph is equal to

$$L = \sum_{i=1}^{8} p(s_i)\ell(s_i) = 2.944$$

where $\ell(s_i)$ represents the mean length of the ways involving s_i. Therefore the mean number of components which must be verified is equal to $2L = 5.888$ but the algorithm shows us the exact way which may be followed.

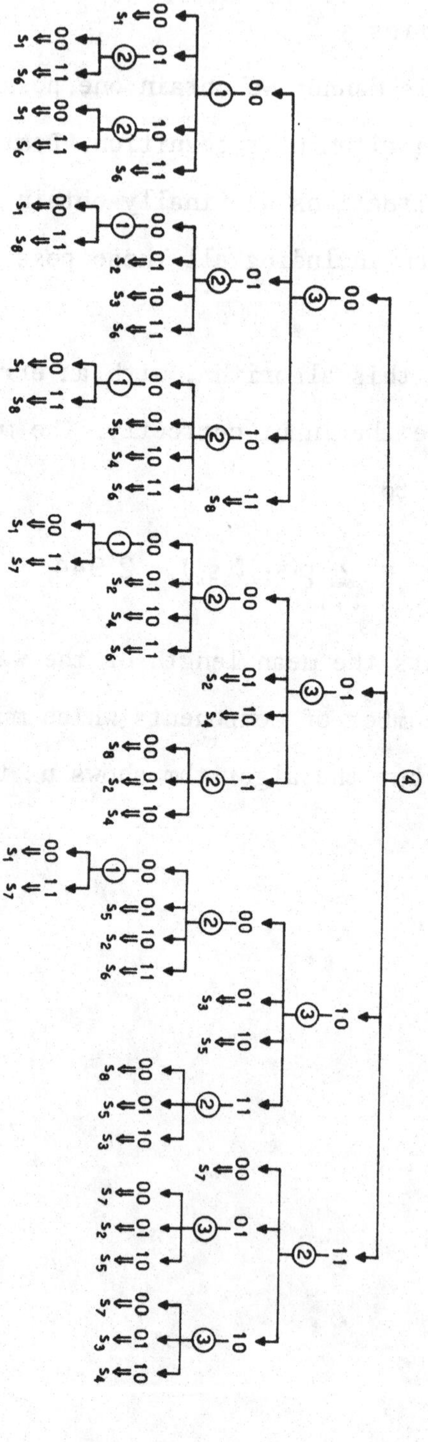

Appendix 1

WEIGHTED ENTROPY [19]

Nowadays there is a marked tendency towards the construction of a qualitative information theory. The weighted entropy is the measure of information supplied by a probabilistic experiment whose elementary events are characterized both by their objective probabilities of occurence and by some qualitative (objective or subjective) weights. The properties, the axiomatic and the maximum value of the weighted entropy are given here. This is the content of one paper which will be published in the Polish journal Reports on Mathematical Physics.

1. INTRODUCTION

The notion of informational entropy has a role of prime importance in statistical physics, and in communication theory. Many problems were clarified by means of this informational entropy as a measure both of the uncertainty and of the information supplied by a probabilistic experiment.

Having its origin in the famous Boltzmann's Function the Shannon's entropy rapidly became a useful tool in several domains, especially in that of communication theory (Shannon, Feinstein, McMillan, Hincin, Kolmogorov), statistical physics (Ingarden, Jaynes), mathematical statistics (Kullbach) and also in measurement theory (Majernìk not to mention other domains such as linguistic, music or social sciences.

Excepting the applications of classical information theory in all these fields, several generalizations of the Shannon entropy were proposed and without any pretensions to completion reference can be made here to Kolmogorov's and Rényi's entropy, α-entropy, Kullbach's relative information, Perez-Csiszar's f-entropy, the Weiss' objective-subjective entropy, or the general axiomatic of information without probability given by Ingarden and Urbanik [22], [13]). Thus the notion of information can be put before that of probability from the point of view of importance and generality.

Underlining the importance of Shannon's entropy, it is at the same time necessary to notice that this formula gives us the measure of information only as function of the probabilities of occurrence of the different events. There are, however, a number of fields dealing with random events where both the probabilities of occurrence and some qualtative characteristics of these events must be taken into account. For instance, in a two-person game it is necessary to take into account both the probabilities of differentvariants of the game (i.e. the random strategies of the players) and the wins corresponding to these variants.

In a physical experiment, it is often very difficult to neglect the subjective aspects related to the various goals of the experiment. At the same time the possible states of a physical system may be very different from the point of view of a given qualitative characteristic. In statistical physics usually all elementary events have the same importance i.e. they are physically equivalent, but this situation is not general. In this last situation it is therefore necessary to associate both the probability of occurrence and the qualitative weight with every elementary event. A criterion for a qualitative differentiation of the possible events of a physical experiment is represented by the relevance, the significance or the utility of the information they carry with respect to a qualitative characteristic. The occurrence of an event removes a double uncertainty, i.e. the quantitative one related to its probability of occurrence and the qualitative one related to its utility for the fulfilment of the goal or to its significance with respect to a given qualitative characteristic.

Of course, the qualitative weight of an event may be independent of its objective probability of occurrence, for instance an event of small probability can havegreat weight likewise an event of great probability can have a very small weight. Naturally, to attach weight to every elementary event is not so easily done. These weights may have either an objective or a subjective character. Thus the weight of one event may express some qualitative objective characteristic but it may express also the subjective utility of the respective event with respect to the experimenter's goal. Also, this weight attached to an elementary event may be related to the subjective probability of occurrence of the respective event, which does not always coincide with the objective probability of occurrence.

We shall suppose that these qualitative weights are non-negative finite real numbers as the usual weights in physics or as the utilities in decision theory. Also, if one event is more relevant, more significant, more useful (with respect to a goal or with respect to a given qualitative point of view) than another event, the weight of the first event will be greater than the corresponding weight of the second one.

How can the amount of information supplied by a probability space, i.e. by a probabilistic experiment, whose elementary events are characterized both by the probabilities of occurrence and by some qualitative (objective or subjective) weights be evaluated? What, in particular is the amount of information supplied by a probabilistic experiment when the probabilities calculated by the experimenter (i.e. the subjective probabilities) do not coincide with the objective probabilities of occurence of these random events?

In the present paper we shall give a formula for the entropy as measure of uncertainty of information supplied by a probabilistic experiment depending both on the probabilities of occurrence and of qualitative (objective or subjective) weights of the possible events. This entropy will be called the weighted entropy. In the following paragraphs the properties, the axiomatic treatment and finally the extremal property of the weighted entropy will be given.

2. Definition and properties of the weighted entropy.

Given a probabilistic physical experiment whose corresponding probability space has a finite number of elementary events $\omega_1,\dots,\omega_n$ with the objective probabilities of occurrence given respectively by the numbers

$$p_k \geqslant 0\,, \quad (k = 1,\dots,n)\,; \qquad \sum_{k=1}^{n} p_k = 1\,.$$

The different elementary events ω_k are more or less relevant depending upon the experimenter's goal or upon some qualitative characteristic of the physical system taken into consideration; that is they have different (objective or subjective) weights. The weight of an event may be either independent of or dependent on its objective probability of occurrence.

In order to distinguish the events $\omega_1,\dots,\omega_n$ of a goal-directed experiment according to their importance with respect to the experimenter's goal, or according to their significance with respect to a given qualitative characteristic of the physical system taken into consideration we will attach a non-negative number $w_k \geqslant 0$ to each event ω_k directly proportional to its importance or significance mentioned above. We shall term w_k the weight of the elementary event ω_k.

We shall call the expression

$$(1.1)\qquad \mathfrak{I}_n = \mathfrak{I}_n(w_1,\ldots,w_n;p_1,\ldots,p_n) = -\sum_{k=1}^{n} w_k p_k \log p_k$$

<u>the weighted entropy</u>

Let us notice briefly some obvious properties of the weighted entropy. The proofs of the first six properties are immediate.

PROPERTY 1:

$$\mathfrak{I}_n(w_1,\ldots,w_n;p_1,\ldots,p_n) \geqslant 0 .$$

PROPERTY 2: If $w_1 = \ldots = w_n = w$ then

$$\mathfrak{I}_n(w_1,\ldots,w_n;p_1,\ldots,p_n) = -w\sum_{k=1}^{n} p_k \log p_k = H_n(p_1,\ldots,p_n)$$

where H_n is the classical Shannon's entropy. (That is unique excepting an arbitrary multiplicative constant).

PROPERTY 3: If $p_{k_0}=1, p_k=0, (k=1,\ldots,n; k\neq k_0)$ then

$$\mathfrak{I}_n(w_1,\ldots,w_n;p_1,\ldots,p_n) = 0$$

whatever the weights $w_1,\ldots,w_n$.

This last property illustrates the obvious fact that an experiment for which only one event is possible does not supply any information. In this case the Shannon's entropy H_n is also equal to zero. Therefore we are only really interested in the probabilistic experiment having two possible events.

PROPERTY 4: If $p_i = 0$, $w_i \neq 0$ for every $i \in I$ and $p_j \neq 0$, $w_j = 0$ for every $j \in J$ where

$$I \cup J = \{1,2,\ldots,n\}, \quad I \cap J = \emptyset$$

then

$$\mathcal{I}_n(w_1,\ldots,w_n;p_1,\ldots,p_n) = 0 .$$

This property illustrates the understood fact that an experiment whose possible results are useless or non-significant and whose useful or significant events are impossible, yields a total information equal to zero even if the corresponding Shannon's entropy $H_n(p_1,\ldots,p_n)$ is different from zero, if the set J has at least two elements. Particularly, when all events have zero weights a total information of $\mathcal{I}_n = 0$ is attained even if the Shannon's entropy H_n is not null i.e. if the condition $0 < p_k < 1$ is satisfied.

PROPERTY 5:

$$\mathcal{I}_{n+1}(w_1,\ldots,w_n,w_{n+1};p_1,\ldots,p_n,0) = \mathcal{I}_n(w_1,\ldots,w_n;p_1,\ldots,p_n)$$

whatever be the weights $w_1,\ldots,w_n,w_{n+1}$ and the complete system of probabilities $p_1,\ldots,p_n$.

PROPERTY 6: For every non-negative real number λ we have

$$\mathcal{I}_n(\lambda w_1,\ldots,\lambda w_n;p_1,\ldots,p_n) = \lambda\mathcal{I}_n(w_1,\ldots,w_n;p_1,\ldots,p_n).$$

Till now we did not impose any restriction on the weights attached to the elementary events of the physical experiment (non-negativity). Let us suppose that the weight of the union of two incompatible events is the mean value of the weights of the respective events, i.e.

$$(1.2) \qquad w(E \cup F) = \frac{p(E)w(E) + p(F)w(F)}{p(E) + p(F)}$$

whatever the incompatible events E, F, where $w(E)$ is the weight of the event E and $p(E)$ is the probability of the same event E. If E and F are complementary events then

$$w(E \cup F) = p(E)w(E) + (1 - p(E))w(F) .$$

PROPERTY 7: If the rule (1.2) for the weights holds then

$$\mathcal{I}_{n+1}(w_1, \dots, w_{n-1}, w', w''; p_1, \dots, p_{n-1}, p', p'') =$$

$$= \mathcal{I}_n(w_1, \dots, w_n; p_1, \dots p_n) + p_n \mathcal{I}_2\left(w', w''; \frac{p'}{p_n}, \frac{p''}{p_n}\right),$$

where

$$w_n = \frac{p'w' + p''w''}{p' + p''}, \quad p_n = p' + p''.$$

PROOF: Taking into account the definition of the weighted entropy and denoting

$$w_n = \frac{p'w' + p''w''}{p' + p''} , \qquad p_n = p' + p''$$

we have

$$\mathfrak{I}_{n+1}(w_1, \ldots, w_{n-1}, w', w''; p_1, \ldots, p_{n-1}, p', p'') =$$

$$= -\sum_{k=1}^{n-1} w_k p_k \log p_k - w'p' \log p' - w''p'' \log p'' =$$

$$= -\sum_{k=1}^{n-1} w_k p_k \log p_k - w_n p_n \log p_n + w_n p_n \log p_n - w'p' \log p' - w''p'' \log p'' =$$

$$= \mathfrak{I}_n(w_1, \ldots, w_n; p_1, \ldots, p_n) + (w'p' + w''p'') \log p_n - w'p' \log p' - w''p'' \log p'' =$$

$$= \mathfrak{I}_n(w_1, \ldots, w_n; p_1, \ldots, p_n) + p_n \left(-w' \frac{p'}{p_n} \log \frac{p'}{p_n} - w'' \frac{p''}{p_n} \log \frac{p''}{p_n} \right) =$$

$$= \mathfrak{I}_n(w_1, \ldots, w_n; p_1, \ldots, p_n) + p_n \mathfrak{I}_2\left(w', w''; \frac{p'}{p_n}, \frac{p''}{p_n} \right) .$$

q.e.d.

Let us consider the following examples

a.) The weighted entropy (1.1) and

$$w_k = -\frac{p_k}{\log p_k} , \quad (k = 1, \ldots, n) . \tag{1.3}$$

In this case the weight of every elementary event

has an objective character representing the ratio between the objective probability of occurrence and the amount of the information supplied by the respective event. In this case we obtain for the weighted entropy the expression

$$\mathcal{I}_n = \sum_{k=1}^{n} p_k^2$$

i.e. Onicescu's informational energy (see [16]) introduced in information theory by analogy to the kinetic energy from mechanics.

b.) Given a probabilistic experiment whose elementary events have the objective probabilities of occurrence $p_1,\dots,p_n$ Let $q_1,\dots,q_n$ the subjective probabilities of occurrence of the same events established by an experimenter, be denoted by $q_1,\dots,q_n$ If we attached to every elementary event the subjective weight

$$w_k = \frac{q_k}{p_k}\,, \quad (k = 1,\dots,n) \tag{1.4}$$

representing the ratio between subjective and objective probability of occurrence of the event ω_k then the expression

$$\mathcal{I}_n = -\sum_{k=1}^{n} q_k \log p_k \tag{1.5}$$

is obtained for the weighted entropy.

If

$$x_k = \frac{p_k}{q_k}, \quad (k = 1,\ldots,n)$$

is inserted in the Jensen's inequality

$$\sum_{k=1}^{n} q_k \log x_k \leq \log\left(\sum_{k=1}^{n} q_k x_k\right)$$

then the inequality

$$\mathfrak{I}_n = -\sum_{k=1}^{n} q_k \log p_k \geq -\sum_{k=1}^{n} q_k \log q_k = H_n \tag{1.6}$$

is obtained showing us that the subjective-objective measure of uncertainty $\mathfrak{I}_n$ in this case is greater than the measure of subjective uncertainty H_n as a consequence of the fact that the subjective probabilities do not coincide with the objective probabilities of occurrence. This means that the degree of uncertainty of the objective probabilities of events is supplemented by another amount of uncertainty as a consequence of the incomplete estimation of these probabilities. In (1.6) we have equality if and only if

$$p_k = q_k, \quad (k = 1,\ldots,n).$$

Let us notice that the weights given by (1.4) satisfy the rule (1.2) while the weights given by (1.3) do not satisfy it. Indeed, according to (1.4) we have

$$w(\omega_i \cup \omega_j) = \frac{q(\omega_i \cup \omega_j)}{p(\omega_i \cup \omega_j)} = \frac{q_i + q_j}{p_i + p_j} = \frac{\frac{q_i}{p_i}p_i + \frac{q_j}{p_j}p_j}{p_i + p_j} = \frac{p_i w_i + p_j w_j}{p_i + p_j}$$

i.e. the inequality (1.2).

We notice here that the rule (1.2) is satisfied by a large kind of weights in physics and also by the utilities in the decision theory according to a von Neumann's well-known axiom. Let us notice also that usually the weights attached to the elementary events are independent of the probabilities of occurrence of these events.

Finally, if all the weights are equal, i.e. if $w_1 = \ldots = w_n = w$ then the rule (1.2) is obviously satisfied and property 7 becomes the well-known property of Shannon's entropy

$$H_{n+1}(p_1, \ldots, p_{n-1}, p', p'') = H_n(p_1, \ldots, p_n) + p_n H_2\left(\frac{p'}{p_n}, \frac{p''}{p_n}\right)$$

where

$$p_n = p' + p'' .$$

3. Axiomatic for the weighted entropy.

We are interested here in the unicity problem of the weighted entropy. In the proof of the unicity theorem that will be given here we shall use the following known lemma:

LEMMA: If $L: N \longrightarrow R^{+}$ is a non-negative function defined on the set of natural numbers and if we have

$$L(mn) = L(m) + L(n) ,$$

$$\lim_{n \to \infty} [L(n) - L(n-1)] = 0$$

then

$$L(n) = \lambda \log n \tag{1.7}$$

where λ is a positive constant.

For the Proof of Rényi see the book [13] .

Also, through this paragraph we shall suppose that the weights attached to the elementary events satisfy the equality (1.2).

Now let us prove the following unicity theorem:

THEOREM 1: Given the sequence of non-negative real-valued funtions

$$(\mathfrak{I}_n(w_1,\ldots,w_n;p_1,\ldots,p_n))_{1\leqslant n\leqslant\infty} ,$$

where every $\mathfrak{I}_n(w_1,\ldots,w_n;p_1,\ldots,p_n)$ is defined on the set

$$w_k \geqslant 0 , \quad p_k \geqslant 0 , \quad (k = 1,\ldots,n) ; \quad \sum_{k=1}^{n} p_k = 1 .$$

Let us suppose that the following four axioms hold:

A_1) $\mathfrak{I}_2(w_1, w_2; p, 1-p)$ is a continuous function of p on the interval $[0,1]$;

A_2) $\mathfrak{I}_n(w_1,\ldots,w_n;p_1,\ldots,p_n)$ is a symmetric function for all pairs variables (w_k,p_k), $(k=1,\ldots,n)$.

A_3) If

$$w_n = \frac{p'w'+p''w''}{p'+p''}, \quad p_n = p'+p'',$$

then

$$\mathfrak{I}_{n+1}(w_1,\ldots,w_{n-1},w',w'';p_1,\ldots,p_{n-1},p',p'') =$$

(1.8)

$$= \mathfrak{I}_n(w_1,\ldots,w_n;p_1,\ldots,p_n) + p_n\mathfrak{I}_2\left(w',w'';\frac{p'}{p_n},\frac{p''}{p_n}\right);$$

A_4) If all the probabilities are equal then

$$\mathfrak{I}_n\left(w_1,\ldots,w_n;\frac{1}{n},\ldots,\frac{1}{n}\right) = L(n)\frac{w_1+\ldots+w_n}{n}$$

$L(n)$ being a positive number for every n;

Then we have

$$\mathfrak{I}_n(w_1,\ldots,w_n;p_1,\ldots,p_n) = -\lambda\sum_{k=1}^{n} w_k p_k \log p_k, \tag{1.9}$$

where λ is an arbitrary positive constant.

The axiom A_3 is simply property 7 mentioned in paragraph § 2. Finally, the last axiom A_4 states that if all the probabilities are equal then the weighted entropy is proportional to the mean value of weights.

PROOF OF THE THEOREM 1: a) From A_3 we have

$$\mathfrak{J}_3\left(w_1,w_2,w_3;\frac{1}{2},\frac{1}{2},0\right) = \mathfrak{J}_2\left(w_1,w_2;\frac{1}{2},\frac{1}{2}\right) + \frac{1}{2}\mathfrak{J}_2(w_2,w_3;1,0) .$$

But A_2 and A_3 imply

$$\mathfrak{J}_3\left(w_1,w_2,w_3;\frac{1}{2},\frac{1}{2},0\right) = \mathfrak{J}_3\left(w_3,w_2,w_1;0,\frac{1}{2},\frac{1}{2}\right) =$$

$$= \mathfrak{J}_2\left(w_3,\frac{w_1+w_2}{2};0,1\right) + \mathfrak{J}_2\left(w_1,w_2;\frac{1}{2},\frac{1}{2}\right) .$$

Then

$$\mathfrak{J}_2(w_2,w_3;1,0) = 2\mathfrak{J}_2\left(\frac{w_1+w_2}{2},w_3,1,0\right),$$

no matter what weights w_1,w_2,w_3. Particularly if $w_1=w_2$ is used

$$\mathfrak{J}_2(w_2,w_3;1,0) = 2\mathfrak{J}_2(w_2,w_3;1,0)$$

is obtained for every w_2,w_3 and therefore,

$$\mathfrak{J}_2(w',w'';1,0) = 0 \tag{1.10}$$

whatever the weights w',w''.

b) Applying A_3 and the equality (1.10) we obtain

$$\mathfrak{J}_{n+1}(w_1,\ldots,w_n,w_{n+1};p_1,\ldots,p_n,0) = \tag{1.11}$$

$$= \mathcal{I}_n(w_1,\dots,w_n;p_1,\dots,p_n) + p_n\mathcal{I}_2(w_n,w_{n+1};1,0) =$$

(1.11) $$\mathcal{I}_n(w_1,\dots,w_n;p_1,\dots,p_n)\ .$$

c) We have also the equality

$$\mathcal{I}_{n+m-1}(w_1,\dots,w_{n-1},w_1',\dots,w_m';p_1,\dots,p_{n-1},p_1',\dots,p_m') =$$

(1.12) $$= \mathcal{I}_n(w_1,\dots,w_n;p_1,\dots,p_n) + p_n\mathcal{I}_m\left(w_1',\dots,w_m';\frac{p_1'}{p_n},\dots,\frac{p_m'}{p_n}\right)$$

where

$$w_n = \frac{p_1'w_1' + \dots + p_m'w_m'}{p_n}\ , \quad p_n = p_1' + \dots + p_m'\ .$$

The above mentioned fact will be proved by induction. Indeed, for $m=2$ we have just the axiom A_3. Let us suppose that the equality (1.12) is verified for m and let us prove its validity for $m+1$. Taking into account the equality (1.11) we may suppose that $p_i' > 0$ for every $i=1,\dots,m$. Then A_3 and (1.11) imply

$$\mathcal{I}_{n+m}(w_1,\dots,w_{n-1},w_1',\dots,w_{m+1}';p_1,\dots,p_{n-1},p_1',\dots,p_{m+1}') =$$

$$= \mathcal{I}_{n+1}(w_1,\dots,w_{n-1},w_1',w'';p_1,\dots,p_{n-1},p_1',p'') +$$

$$+ p''\mathcal{I}_m\left(w_2',\dots,w_{m+1}';\frac{p_2'}{p''},\dots,\frac{p_{m+1}'}{p''}\right) = \mathcal{I}_n(w_1,\dots,w_n;p_1,\dots,p_n) +$$

$$+ p_n\mathcal{I}_2\left(w_1',w'';\frac{p_1'}{p_n},\frac{p''}{p_n}\right) + p''\mathcal{I}_m\left(w_2',\dots,w_{m+1}';\frac{p_2'}{p''},\dots,\frac{p_{m+1}'}{p''}\right)$$

where

$$w'' = \frac{p_2' w_2' + \ldots + p_{m+1}' w_{m+1}'}{p''} \quad , \quad p'' = p_2' + \ldots + p_{m+1}' \, ,$$

and

$$w_n = \frac{p_1' w_1' + p'' w''}{p_1' + p''} = \frac{p_1' w_1' + p_2' w_2' + \ldots + p_{m+1}' w_{m+1}'}{p_1' + p_2' + \ldots + p_{m+1}'} = \frac{p_1' w_1' + \ldots + p_{m+1}' w_{m+1}'}{p_n}$$

$$p_n = p_1' + \ldots + p_{m+1}' \, .$$

Assuming that (1.12) is true for m fixed, we obtain

$$p_n \mathfrak{I}_{m+1}\left(w_1', \ldots, w_{m+1}'; \frac{p_1'}{p_n}, \ldots, \frac{p_{m+1}'}{p_n}\right) = p_n \mathfrak{I}_2\left(w_1', w''; \frac{p_1'}{p_n}, \frac{p''}{p_n}\right) + $$
$$+ p_n \frac{p''}{p_n} \mathfrak{I}_m\left(w_2', \ldots, w_{m+1}'; \frac{p_2'}{p''}, \ldots, \frac{p_{m+1}'}{p''}\right) \, . \qquad (1.14)$$

From (1.13) and (1.14) we have

$$\mathfrak{I}_{n+m}(w_1, \ldots, w_{n-1}, w_1', \ldots, w_{m+1}'; p_1, \ldots, p_{n-1}, p_1', \ldots, p_{m+1}') =$$
$$= \mathfrak{I}_n(w_1, \ldots, w_n; p_1, \ldots, p_n) + p_n \mathfrak{I}_{m+1}\left(w_1', \ldots, w_{m+1}'; \frac{p_1'}{p_n}, \ldots, \frac{p_{m+1}'}{p_n}\right) .$$

Therefore the equality (1.12) is true for $m+1$ and hence for arbitrary m .

d) When the eq. (1.12) is applied several times, the following result is obtained

$$\mathfrak{I}_{m_1+\ldots+m_n}(w'_{11},\ldots,w'_{1m_1},\ldots,w'_{n1},\ldots,w'_{nm_n};p'_{11},\ldots,p'_{1m_1},\ldots,p'_{n1},\ldots,p'_{nm_n})=$$

$$=\mathfrak{I}_n(w_1,\ldots,w_n;p_1,\ldots,p_n)+\sum_{i=1}^{n}p_i\,\mathfrak{I}_{m_i}\left(w'_{i1},\ldots,w'_{im_i};\frac{p'_{i1}}{p_i},\ldots,\frac{p'_{im_i}}{p_i}\right)$$

(1.15)

where

$$p_i = p'_{i1}+\ldots+p'_{im_i} > 0\ , \quad (i = 1,\ldots,n)$$

$$w_i = \frac{p'_{i1}w'_{i1}+\ldots+p'_{im_i}w'_{im_i}}{p_i}\ .$$

e) When this last equality (1.15) for $m_1=\ldots=m_n=m$ is applied, the following result is obtained

$$\mathfrak{I}_{mn}(w'_{11},\ldots,w'_{1m},\ldots,w'_{n1},\ldots,w'_{nm};p'_{11},\ldots,p'_{1m},\ldots,p'_{n1},\ldots,p'_{nm}) =$$

$$= \mathfrak{I}_n(w_1,\ldots,w_n;p_1,\ldots,p_n)+\sum_{i=1}^{n}p_i\,\mathfrak{I}_m\left(w'_{i1},\ldots,w'_{im};\frac{p'_{i1}}{p_i},\ldots,\frac{p'_{im}}{p_i}\right)$$

If we take

$$p'_{ij} = \frac{1}{mn}\ , \quad (i = 1,\ldots,n\ ;\ j = 1,\ldots,m),$$

we have

$$p_i = \frac{1}{n}\ , \quad (i = 1,\ldots,n)\ ,$$

and therefore

$$\mathfrak{I}_{mn}\left(w'_{1_1},\dots,w'_{1_m},\dots,w'_{n_1},\dots,w'_{n_m},\frac{1}{mn},\dots,\frac{1}{mn}\right) =$$

$$= \mathfrak{I}_n\left(\frac{w'_{1_1}+\dots+w'_{1m}}{m},\dots,\frac{w'_{n_1}+\dots+w'_{nm}}{m};\frac{1}{n},\dots,\frac{1}{n}\right) +$$

$$+ \sum_{i=1}^{n}\frac{1}{n}\mathfrak{I}_m\left(w'_{i_1},\dots,w'_{im};\frac{1}{m},\dots,\frac{1}{m}\right).$$

Taking into account the axiom A_4 we obtain

$$L(mn)\frac{1}{mn}\sum_{i=1}^{n}(w'_{i_1}+\dots+w'_{im}) =$$

$$= L(n)\frac{1}{n}\sum_{i=1}^{n}\frac{w'_{i_1}+\dots+w'_{im}}{m} + L(m)\frac{1}{n}\sum_{i=1}^{n}\frac{w'_{i_1}+\dots+w'_{im}}{m},$$

or

$$L(mn) = L(n) + L(m). \tag{1.16}$$

f) Using the equality (1.12)

$$\mathfrak{I}_n\left(w_1,\dots,w_n;\frac{1}{n},\dots,\frac{1}{n}\right) = \mathfrak{I}_2\left(w_1,\frac{w_2+\dots+w_n}{n-1};\frac{1}{n},\frac{n-1}{n}\right) +$$

$$+ \frac{n-1}{n}\mathfrak{I}_{n-1}\left(w_2,\dots,w_n;\frac{1}{n-1},\dots,\frac{1}{n-1}\right).$$

Applying the axiom A_4

$$L(n)\frac{w_1 + \ldots + w_n}{n} = \Im_2\left(w_1, \frac{w_2 + \ldots + w_n}{n-1}; \frac{1}{n}, \frac{n-1}{n}\right) +$$

$$+ \frac{n-1}{n}\frac{w_2 + \ldots + w_n}{n-1} L(n-1).$$

for every $w_1, w_2, \ldots, w_n$. Let $w_1 = 0$. Then

$$\Im_2\left(0, \frac{w_2 + \ldots + w_n}{n-1}; \frac{1}{n}, \frac{n-1}{n}\right) = \frac{w_2 + \ldots + w_n}{n-1}\left[L(n) - L(n-1)\right],$$

whichever be $w_2, \ldots, w_n$. Let us take $w_1 = , \ldots, = w_n = w$. We have

$$0 \leqslant \Im_2\left(0, w; \frac{1}{n}, \frac{n-1}{n}\right) = \frac{n-1}{n} w \left[L(n) - L(n-1)\right],$$

and according to the axiom A_1 and to the equality (1.10) we obtain

$$0 = \Im_2(0, w; 0, 1) = \lim_{n \to \infty} \Im_2\left(0, w; \frac{1}{n}, \frac{n-1}{n}\right) =$$

$$= \lim_{n \to \infty} \frac{n-1}{n} w\left[L(n) - L(n-1)\right] = w \lim_{n \to \infty}\left[L(n) - L(n-1)\right],$$

for every non-negative real number w, i.e.

(1.17) $$\lim_{n \to \infty}\left[L(n) - L(n-1)\right] = 0.$$

g) From (1.16) and (1.17) the lemma mentioned at the beginning of this paragraph implies

$$L(n) = \lambda \log n \,, \tag{1.18}$$

where λ is an arbitrary positive constant.

h) $n=2$, $m_1 = r$, $m_2 = s-r$, $p'_{ij} = \frac{1}{s}$ in the equality (1.15)

Then

$$p_1 = p'_{11} + \ldots + p'_{1r} = \frac{r}{s} \,,$$

$$p_2 = p'_{21} + \ldots + p'_{2,s-r} = \frac{s-r}{s} \,,$$

$$w_1 = \frac{p'_{11}w'_{11} + \ldots + p'_{1r}w'_{1r}}{p'_{11} + \ldots + p'_{1r}} = \frac{1}{r}\sum_{i=1}^{r} w'_{1i} \,,$$

$$w_2 = \frac{p'_{21}w'_{21} + \ldots + p'_{2,s-r}w'_{2,s-r}}{p'_{21} + \ldots + p'_{2,s-r}} = \frac{1}{s-r}\sum_{j=1}^{s-r} w'_{2j} \,.$$

Taking into account all these values (1.15) yields

$$\mathcal{I}_s\left(w'_{11}, \ldots, w'_{1r}, w'_{21}, \ldots, w'_{2,s-r}; \frac{1}{s}, \ldots, \frac{1}{s}\right) =$$

$$= \mathcal{I}_2(w_1, w_2; p_1, p_2) + p_1\mathcal{I}_r\left(w'_{11}, \ldots, w'_{1r}; \frac{1}{r}, \ldots, \frac{1}{r}\right) + p_2\mathcal{I}_{s-r}\left(w'_{21}, \ldots, w'_{2,s-r}; \frac{1}{s-r}, \ldots, \frac{1}{s-r}\right). \tag{1.19}$$

But according to A_4 and (1.18)

$$\mathfrak{I}_s\left(w'_{11},\dots,w'_{1r},w'_{21},\dots,w'_{2,s-r};\frac{1}{s},\dots,\frac{1}{s}\right) = \lambda\frac{1}{s}\left(\sum_{i=1}^{r} w'_{1i} + \sum_{j=1}^{s-r} w'_{2j}\right)\log s\,,$$

$$\mathfrak{I}_r\left(w'_{11},\dots,w'_{1r};\frac{1}{r},\dots,\frac{1}{r}\right) = \lambda\frac{1}{r}\sum_{i=1}^{r} w'_{1i}\log r\,,$$

$$\mathfrak{I}_{s-r}\left(w'_{21},\dots,w'_{2,s-r};\frac{1}{s-r},\dots,\frac{1}{s-r}\right) = \lambda\frac{1}{s-r}\sum_{j=1}^{s-r} w'_{2j}\log(s-r)\,,$$

and from (1.19)

$$\mathfrak{I}_2(w_1,w_2;p_1,p_2) = \lambda\frac{1}{s}\left(\sum_{i=1}^{r} w'_{1i} + \sum_{j=1}^{s-r} w'_{2j}\right)\log s -$$

$$-\lambda\frac{r}{s}\frac{1}{r}\sum_{i=1}^{r} w'_{1i}\log r - \lambda\frac{s-r}{s}\frac{1}{s-r}\sum_{j=1}^{s-r} w'_{2j}\log(s-r) =$$

$$= -\lambda\frac{r}{s}\left(\frac{1}{r}\sum_{i=1}^{r} w'_{1i}\right)\log\frac{r}{s} - \lambda\frac{s-r}{s}\left(\frac{1}{s-r}\sum_{j=1}^{s-r} w'_{2j}\right)\log\frac{s-r}{s} =$$

$$= -\lambda w_1 p_1\log p_1 - \lambda w_2 p_2\log p_2\,.$$

e) The formula (1.9) for $n=2$ has been proved at the point h). Assuming that it is true also for n we shall see that (1.9) holds for $n+1$ too. Therefore from (1.8) gives the following result

$$\mathfrak{I}_{n+1}(w_1,\dots,w_{n-1},w',w'';p_1,\dots,p_{n-1},p',p'') =$$

$$= -\lambda\sum_{i=1}^{n} w_i p_i\log p_i - p_n\left(\lambda w'\frac{p'}{p_n}\log\frac{p'}{p_n} + \lambda w''\frac{p''}{p_n}\log\frac{p''}{p_n}\right) =$$

$$= -\lambda \sum_{i=1}^{n-1} w_i p_i \log p_i - \lambda w_n p_n \log p_n - \lambda w' p' \log p' - \lambda w'' p'' \log p'' +$$

$$+ \lambda w' p' \log p_n + \lambda w'' p'' \log p_n = -\lambda w_1 p_1 \log p_1 - \ldots - \lambda w_{n-1} p_{n-1} \log p_{n-1} -$$

$$- \lambda w' p' \log p' - \lambda w'' p'' \log p'' ,$$

because

$$w' p' + w'' p'' = w_n p_n .$$

Therefore the equality (3.9) is true for arbitrary n . q.e.d.

Remark: Let us take $w_1 = \ldots = w_n = 1$. Then, as we have already seen, the weighted entropy is simply the Shannon's entropy. At the same time, in this case the axiom A_4 is obviously satisfied, becoming

$$\mathfrak{I}_n\left(1, \ldots, 1; \frac{1}{n}, \ldots, \frac{1}{n}\right) = H_n\left(\frac{1}{n}, \ldots, \frac{1}{n}\right) = L(n) ,$$

and A_1, A_2, A_3 are simply Faddeev's well-known axioms for the Shannon's entropy (see [13]):

A_1) $H_2(p, 1-p)$ is a continuous function of $p \in [0,1]$

A_2) $H_n(p_1, \ldots, p_n)$ is a symmetric function of all variables p_i

A_3)

$$H_{n+1}(p_1,\dots,p_{n-1},p',p'') = H_n(p_1,\dots,p_n) + p_n H_2\left(\frac{p'}{p_n},\frac{p''}{p_n}\right),$$

where

$$p_n = p' + p''.$$

4. Principle of maximum information.

The principle of maximum information established independently by Ingarden , Jaynes and Kullbach has many important applications in statistical physics and mathematical statistics. Many important random distributions may be obtained as consequences of this principle. Taking into account some constraints it is possible to obtain by maximization of information the equal probability distribution (the finite discrete case with no constraints), the geometrical distribution (continuous or discrete case when the constraint is the mean value of some fixed random variable), the Gaussian distribution (continuous case the constraints being the mean value on the variance of some fixed random variable). Recently, Ingarden and Kossakowski even obtained the Poisson distribution in the countable discrete case by maximizing the relative information, the constraint being the mean value of the random variable $f(\omega_n) = n$ $(n = 0,1,2,\dots)$ and the initial measure being

$$m(\omega_n) = \frac{1}{n!}, \quad (n = 0,1,2,\ldots).$$

Let us now give the expression of the probability distribution maximizing the weighted entropy. Dealing with natural logarithms we shall prove the following theorem (principle of maximum information with no constraint):

THEOREM 2: Let there be the probability distribution

$$p_i > 0, \quad (i = 1,\ldots,n); \quad \sum_{i=1}^{n} p_i = 1 \tag{1.20}$$

and the weights $w_i \geq 0$, $(i = 1,\ldots,n)$. The weighted entropy

$$\mathcal{J}_n = \mathcal{J}_n(w_1,\ldots,w_n; p_1,\ldots,p_n) = -\sum_{i=1}^{n} w_i p_i \log p_i$$

is maximum if and only if

$$p_i = e^{-\frac{\alpha}{u_i}-1} \quad (i = 1,\ldots,n)$$

where α is the solution of the equation

$$\sum_{i=1}^{n} e^{-\frac{\alpha}{u_i}-1} = 1.$$

PROOF: Because $\log x < x-1$ for every $x-1$ and $\log x = x-1$ if and only if $x=1$, by using the Lagrange's multipliers method we obtain $H(-x\log x \leq e^{-1})$

$$\mathcal{I}_n - \alpha = \sum_{i=1}^{n} w_i p_i \log \frac{1}{p_i} - \alpha \sum_{i=1}^{n} p_i = \sum_{i=1}^{n} p_i \left(w_i \log \frac{1}{p_i} - \alpha \right) =$$

$$= \sum_{i=1}^{n} p_i \log \left(\frac{1}{p_i^{w_i}} e^{-\alpha} \right) \leq \sum_{i=1}^{n} w_i e^{-\frac{\alpha}{w_i}-1} .$$

The equation holds if and only if

$$p_i = e^{-\frac{\alpha}{w_i}-1} \quad (i = 1,\dots,n)$$

These probabilities must veryfy the relation (1.20) i.e.

$$\sum_{i=1}^{n} e^{-\frac{\alpha}{w_i}-1} = 1$$

Remark 1: If all events have the same weight $w_1 = \dots = w_n = 1$ then

$$p_i = \frac{1}{n} \quad (i = 1,\dots,n)$$

i.e. we obtain the equal probability distribution.

Remark 2: The definition together with some properties of the weighted entropy were given briefly in the paper

[12]. The axiomatic i.e. the whole § 3 and the essential property 7 of the weighted entropy are presented here for the first time.

Remark 3: In paragraph § 4 we did not suppose that the weights satisfy the rule (1.2).

Remark 4: During the author's visit in Budapest, Prof. I. Vincze suggested that the relative weighted entropy

$$\log n \sum_{i=1}^{n} w_i p_i - \mathcal{I}_n = \sum_{i=1}^{n} w_i p_i \log n p_i$$

be considered instead of the weighted entropy which permits a natural passage to the continuous case

Appendix 2

MODIFICATION OF RANDOM STRATEGIES IN NON–ZERO SUM GAMES [18]

Only a few features relating to the game theory will be presented here, thus gaving a partial answer to the recent objections concerning the topics of classical game theory. The game is an important cybernetic system closely connected to the decision theory. It is a well-developed topic of cybernetics but certain objections could not fail to appear. Thus n -person games, with n larger than 2, are highly significant in the mathematical theory of conflict situations, the presence and attitude of a third player assuming overwhelming importance.

There are also games with neither definite duration nor stable utilities for the player. These may far more correctly be assessed as repeated games in which one player, at a given moment may alter his strategy or utilities (profitable or detrimental to other players), thus causing changes in the strategies or utilities of those affected by the first change.

One should likewise remember the fact that game theory proper studies the games "from outside" from the angle of an umpire who is not involved in the game.

Reference will be made in the following to a three person game. The leap form two to three persons seems to be essential, while generalization of results from three to $n>3$ persons will only entail difficulties of writing. Suppose that these three players are engaged in a game. Each of them is acting on his own, with a definite random strategy and a definite mean payoff, the latter being determined by the random strategies of all the players and by the utilities of the various possible variants of the game. At a given moment, one player (player 1) introduces surprise, or, more exactly, alters his random strategy by adopting another, more indeterminate to the other players, thus bringing surprise into the actions. Surprise not only changes the existing situation (the players' mean payoffs being affected in consequence), but is disagreable in itself since the actions of the player who introduced it have become much more indeterminate, much more unpredictable, giving rise to a feeling of panic. Player 2, therefore is faced with surprise, and wants to react. But how will he do it? What should he take into account? In the first place he would like , by merely altering his own random strategy, to act, in response to the change, so that the mean payoff of player 1 (who introduced the element of surprise) should be smaller than the one that player obtained before the introduction of surprise, which thus becomes unprofitable even to player 1, who resorted to it. (In fact, player 1 by introducing surprise also intended to confuse the others, in addition to increasing his own mean payoff through this change of random strategy; but the response of player 2, when possible, results in making this mean payoff, which has risen through the change in the strategy of player 1, become smaller than the original mean payoff in consequence of the change in the random strategy of player 2, (who has responded).

Secondly, player 2, when responding to surprise by changing his random strategy, must also consider his own interest. He will try to alter his own random strategy so as to prevent, despite the surprise introduced by player 1 (i.e. when confronted with a new random strategy of player 1), his own mean payoff being less than before the introduction of surprise.

Lastly, in the third place, player 2 must not, by changing his random strategy, affect the mean payoff of player 3, who is in no way involved in the introduction of surprise. Consequently, player 2 will modify his own random strategy so that the mean payoff of player 3 (a neutral, for the time being!), which may or may not be affected by the surprise introduced by player 1, is not diminished as a result of this modification.

Owing to the feedback characteristic of any game, modification of the random strategy of one player will influence the mean payoff of the others. Confronted with the surprise element introduced by player 1, player 2, wants to respond, but his response must take account of the three above-mentioned factors

1) It should result in a smaller mean payoff for player 1 than before the introduction of surprise (which thus becomes unprofitable for its originator; 2) It should counteract the possible negative effects of surprise on his own mean payoff

(i.e. neutralization of the negative effects of surprise on one's own interest); 3) It should not affect negatively the mean pay-off of player 3 (i.e. the response should not be injurious to the neutral's interests).

We shall prove the existence of a wide class of games in which, by merely changing his own random strategy, player 2 can give the correct answer (i.e. the one conformable to the three aspects mentioned above) to the introduction of the element of surprise. These will be A type games.

The class of games in which correct response is possible becomes wider if player 2 changes, in addition to his random strategy, his utilities of the various possible variants of the game. These will be the B type games.

It should be noted that in games of the A and B types, player 2 will manage without assistance from player 3 who, despite the introduction of surprise, continues to be a neutral and will not change his strategy and utilities, thus calmly accepting the consequences of surprise.

The renunciation of a position of neutrality by player 3 will lead to a corresponding growth in the importance of his own role. By changing his own strategy and, possibly, his own utilities, player 3 can almost always tip the balance, either by reacting in his turn to the surprise introduced by player 1 (thus backing the response of player 2), or by reducing (or annulling) the response of player 2 (thus backing the surprise introduced by player 1).

The class of games in which correct response by player 2 is possible will grow wider if player 2 changes his own random strategy and utilities, and if, at the same time, player 3, by changing his random strategy, and, possibly, his utilities, cooperates with player 2. These will be the C type games.

Two conclusions arise notably from the mathematical formation to be developed below. First, the fact that the element of surprise loses its absolute character. Second, the following formulae will indicate the actual manner in which player 2 (in games of the A, B and C types) and player 3 (in games of the C type) must act in order to respond correctly to the introduction of surprise.

What underlies the mathematical formalism which has led us to these conclusions? Before replaying to this question let us go back to the objections raised earlier as to the current methods and results of the theory of games. We pointed out that surprise in actions is not generally considered in the approaches to this field, although it is fairly frequent in the practice of international relations. In order to tackle it one must resort to different apparatus placed outside game theory proper. This is the apparatus of the mathematical theory of information, more exactly calculation by means of informational entropy and the variational problems connected with entropy. Surprise in actions means a higher amount of indetermination for the other players. The more indetermination is contained in a player's random strategy, the greater the surprise, and the uncertainty of his actions to the other persons. In information theory, however, a measure has been devised for the degree of indetermination contained in a probability distribution. This measure is Shannon's entropy, introduced in 1948, by analogy with Boltzmann's physical entropy, in Shannon's well-known memorandum on the mathematical theory of information. Information theory has developed considerably since 1948. Its apparatus has proved fertile in the study of international relations as well, by providing an approach, (within the theory games) to surprise in actions and its neutralization.

The mathematical apparatus relating to entropy will be employed not only to tackle surprise in actions, but also to obtain a global characterization of the random strategies of the players. We have already mentioned, when dealing with the various types of games which may be interesting in point of a perfect response, the change in random strategy and the suitable alteration of utilities. The suitable alteration of the utilities of the possible variants of the game is much easier (utilities are non-negative real numbers, not subject to any other restrictive condition) but also far more artificial than the alteration of the random strategy itself. It is however of practical use, to find out precisely, while the utilities are kept unchanged, how one should change ones random strategy to ensure a perfect response.

This is a difficult problem because a random strategy is a probability distribution (non-negative numbers subject to the restrictive condition that their sum should equal to 1), and when replaced by another random strategy, i.e. another probability distribution, the diminished probabilities of some actions, i.e. of some possible pure strategies (which diminution is sometimes self-evident) will result, precisely because of the obligatory restrictive relation (the sum of the probabilities composing the random strategy is 1) in increased possibilities for other possible pure strategies. It is therefore very difficult, if not impossible, to state, in many cases, how one should select one's adequate random strategies, or what should characterize the adequate random strategy (or random strategies, as there may be several). Entropy, as the overall characteristic of a probability distribution will provide a solution here, too. There answers are as follows: Suitable distribution should have its entropy (i.e. its degree of indetermination) contained between certain numerical limits. As there are tables for calculating the entropy, it is very easy to find out, through the entropy, the distribution, or distributions, that will solve the problem in hand.

We shall do nothing more than translate into mathematical quantitative terms what has been said above. We shall introduce notations, mathematical characterizations, calculation procedures, examples. Proofs have been removed from the paper and included in two appendices. Appendix 2.1 contains the lemmas (i.e. the general results from information theory we have used) with their proofs; Appendix 2.2 contains the proofs of the mathematical statements made in the paper.

Consider a three-person game, in which the first player adopts the random strategy $(\xi_i)_{1 \leq i \leq r}$ the second adopts the random strategy $(\eta_j)_{1 \leq j \leq s}$ and the third adopts the random strategy $(\zeta_k)_{1 \leq k \leq t}$ where

$$\xi_i \geq 0\,, \quad \eta_j \geq 0\,, \quad \zeta_k \geq 0\,, \quad \sum_i \xi_i = \sum_j \eta_j = \sum_k \zeta_k = 1\,. \tag{2.1}$$

Here and in the following, index i varies from 1 to r, index j varies from 1 to s, and index k varies from 1 to t, the natural numbers r, s and t representing the number of the three players' pure strategies.

Let u^q_{ijk} be the utility for player q $(q=1,2,3)$ of the game variant formed of pure strategies i,j,k of the three players. These utilities are positive real numbers, i.e.

$$u^q_{ijk} \geq 0\,, \quad (i=1,\ldots,r\,;\, j=1,\ldots,s\,;\, k=1,\ldots,t) \qquad (q=1,2,3\,).$$

Since this is a game with independent strategies, the mean payoff of player q $(q=1,2,3)$ will be

$$\mathcal{U}^q(\xi,\eta,\zeta) = \sum_{i,j,k} u^q_{ijk}\,\xi_i\,\eta_j\,\zeta_k\,, \tag{2.2}$$

where summation is done according to all the values of i, j and k and where ξ, η, ζ are the vectors of components

$$\xi = (\xi_1, \dots, \xi_r) \ , \quad \eta = (\eta_1, \dots, \eta_s) \ , \quad \zeta = (\zeta_1, \dots, \zeta_t) \ , \tag{2.3}$$

i.e. the random strategies (2.1) adopted by the players during the game. Each player's actions, or rather, each player's random strategy will contain a certain amount of uncertainty, the indetermination of the probability distribution composing the respective random strategy, measured by Shannon's informational entropy. Thus, the random strategies (2.3) adopted by three players contain the respective indeterminations

$$H(\xi) = -\sum_i \xi_i \log \xi_i \ , \quad H(\eta) = -\sum_j \eta_j \log \eta_j \ ,$$
$$H(\zeta) = -\sum_k \zeta_k \log \zeta_k \tag{2.4}$$

where logarithms are taken in base e. Selection of the logarithm base is not essential since entropy, as the measure of the degree of indetermination is unique apart from an arbitrary positive multiplicative constant which enables us to change the logarithm base.

Suppose that in our repeated game, at a certain moment the first player changes his random strategy so that it contains more uncertainty, and more indetermination for the other players. Surprise means the modification of the random strategy by one player at a certain moment of the game. Player 1 passes from the random strategy ξ to the random strategy

$$\xi^0 = (\xi_1^0, \dots, \xi_r^0), \quad \xi_i^0 \geq 0, \quad \sum_{i=1}^{r} \xi_i^0 = 1. \qquad (2.5)$$

Often the new strategy of player 1 will contain a higher degree of indetermination, i.e.

$$H(\xi^0) > H(\xi) \qquad (2.6)$$

or

$$-\sum_{i=1}^{r} \xi_i^0 \log \xi_i^0 > -\sum_{i=1}^{r} \xi_i \log \xi_i \qquad (2.6')$$

but this last condition is not obligatory. More often than not the new random strategy ξ^0 adopted by the player 1, brings him a greater mean payoff, i.e.

$$u^1(\xi^0, \eta, \zeta) > u^1(\xi, \eta, \zeta) \qquad (2.7)$$

In the following, however we shall not suppose that (2.7) is automatically satisfied, in order to avoid excluding from our consideration those cases when player 1 adopts a random strategy which may not necessarily bring him a greater mean payoff.

Faced with surprise player 2 would like to respond correctly, i.e. to know how to choose a random strategy

$$\begin{aligned} \eta^0 &= (\eta_1^0, \dots, \eta_s^0) \\ \eta_j^0 \geq 0, &\quad \sum_{j=1}^{s} \eta_j^0 = 1 \end{aligned} \qquad (2.8)$$

in place of random strategy η , so that: 1) the mean payoff of the first player, who introduced surprise, will become less than before the introduction of surprise, i.e.

(2.9) $$u^1(\xi^0,\eta^0,\zeta) < u^1(\xi,\eta,\zeta),$$

or, differently written,

(2.9') $$\sum_{i,j,k} u^1_{ijk}\xi^0_i\eta^0_j\zeta_k < \sum_{i,j,k} u^1_{ijk}\xi_i\eta_j\zeta_k ;$$

2) his own mean payoff will not be affected negatively by surprise, i.e.

(2.10) $$u^2(\xi^0,\eta^0,\zeta) \geqslant u^2(\xi,\eta,\zeta)$$

or, differently written,

(2.10') $$\sum_{i,j,k} u^2_{ijk}\xi^0_i\eta^0_j\zeta_k \geqslant \sum_{i,j,k} u^2_{ijk}\xi_i\eta_j\zeta_k ;$$

3) adoption of the new strategy η^0 will not reduce the mean payoff of the player 3, who is neutral, i.e.

(2.11) $$u^3(\xi^0,\eta^0,\zeta) \geqslant u^3(\xi^0,\eta,\zeta),$$

or, explicitly,

(2.11') $$\sum_{i,j,k} u^3_{ijk}\xi^0_i\eta^0_j\zeta_k \geqslant \sum_{i,j,k} u^3_{ijk}\xi^0_i\eta_j\zeta_k .$$

A game is of type A if there is a random strategy η^0, for which the inequalities (2.9), (2.10), (2.11) are simultaneously satisfied, where ξ^0 is a random strategy for which (2.6) will occur or not.

If a game is not of type A, but there are some utilities u^{*2}_{ijk} $(i=1,\dots,r; j=1,\dots,s; k=1,\dots,t)$ and a random strategy also noted so that the inequalities (2.9), (2.11) and

$$u^{*2}(\xi^0,\eta^0,\zeta) \geq u^2(\xi,\eta,\zeta) \tag{2.12}$$

where

$$u^{*2}(\xi^0,\eta^0,\zeta) = \sum_{i,j,k} u^{*2}_{ijk}\,\xi^0_i\,\eta^0_j\,\zeta_k \tag{2.13}$$

are satisfied, that game is considered as a B type game.

Lastly, if a game falls into neither of the above mentioned types but the utilities u^{*2}_{ijk}, u^{*3}_{ijk} $(i=1,\dots,r; j=1,\dots,s; k=1,\dots,t)$ exist together with the random strategies η^0 and ζ^0, so that the inequalities below

$$u^1(\xi^0,\eta^0,\zeta^0) < u^1(\xi,\eta,\zeta) \tag{2.14}$$

$$u^{*2}(\xi^0,\eta^0,\zeta^0) \leq u^2(\xi,\eta,\zeta) \tag{2.15}$$

$$u^{*3}(\xi^0,\eta^0,\zeta^0) \geq u^3(\xi,\eta,\zeta) \tag{2.16}$$

are satisfied simultaneously, the game is of type C, where

$$\zeta^0 = (\zeta_1^0,\dots,\zeta_t^0), \quad \zeta_k^0 \geq 0, \quad \sum_k \zeta_k^0 = 1,$$

(2.17)
$$\mathcal{U}^1(\xi^0,\eta^0,\zeta^0) = \sum_{i,j,k} u^1_{ijk}\,\xi_i^0\eta_j^0\zeta_k^0 ;$$
$$\mathcal{U}^{*2}(\xi^0,\eta^0,\zeta^0) = \sum_{i,j,k} u^{*2}_{ijk}\,\xi_i^0\eta_j^0\zeta_k^0 ;$$
$$\mathcal{U}^{*3}(\xi^0,\eta^0,\zeta^0) = \sum_{i,j,k} u^{*3}_{ijk}\,\xi_i^0\eta_j^0\zeta_k^0 .$$

Obviously, the class of A type games is contained in the class of B type games which, in its turn, is contained in the class of C type games.

Consider now an A type game. The question naturally arises as to: how one should choose the random strategy η^0 for which the inequalities (2.9), (2.10), (2.11) are satisfied simultaneously. Before replying, we shall introduce one more useful notation. Given a probability distribution

$$p_j \geq 0, \quad \sum_{j=1}^{n} p_j = 1,$$

and some non-negative real numbers $u_j \geq 0$ $(j=1,\dots,n)$ we shall call the expression

(2.18)
$$\mathcal{H}(u,p) = -\sum_{j=1}^{n} u_j p_j \log p_j ,$$

weighted entropy attached to the distribution $(p_j)_{1\leq j\leq n}$ with the weighted $u=(u_1,\ldots,u_n)$ where p is the vector $p=(p_1,\ldots,p_n)$. Obviously, the entropy weighted by the vector $1=(1,\ldots,1)$ will coincide with the entropy of the distribution p , i.e.

$$\mathfrak{I}(1,p) = H(p) . \tag{2.19}$$

Using the mathematical apparatus connected with entropy (which is shown in detail in Appendices 1 and 2.2) we shall obtain the following mathematical characterization of random strategy η^0 which assures correct response in an A type game. Noting with

$$\mathcal{U}^q_j(\xi^0,\zeta) = \sum_{i,k} u^q_{ijk}\xi^0_i\zeta_k , \quad (q = 1,2,3;\ j = 1,\ldots,s) \tag{2.20}$$

and with $U^1(\xi^0,\zeta)$ the vector of components

$$U^1(\xi^0,\zeta) = (\mathcal{U}^1_1(\xi^0,\zeta),\ldots,\mathcal{U}^1_s(\xi^0,\zeta)) \tag{2.21}$$

the random strategy η^0 which realizes correct response in an A type game must be so chosen that its entropy $H(\eta^0)$ and its entropy weighted by $U^1(\xi^0,\zeta)$ i.e. $\mathfrak{I}(U^1(\xi^0,\zeta),\eta^0)$ should satisfy the inequalities

$$H(\eta^0) \geq \mathcal{U}^2(\xi,\eta,\zeta) + \log\left(\sum_j e^{-\mathcal{U}^2_j(\xi^0,\zeta)}\right), \tag{2.22}$$

(2.23) $$H(\eta^0) \geq \mathcal{U}^3(\xi^0,\eta,\zeta) + \log\left(\sum_j e^{-\mathcal{U}_j^3(\xi^0,\zeta)}\right),$$

(2.24) $$\mathcal{H}(U^1(\xi^0,\zeta),\eta^0) \geq \sum_j \mathcal{U}_j^1(\xi^0,\zeta) - \mathcal{U}^1(\xi,\eta,\zeta).$$

Under what circumstances will there be at least one solution η^0 for which the entropy and weighted entropy satisfy the inequalities (2.22), (2.23), (2.24)? The wider conditions under which there is a solution η^0 ensuring correct response will be obtained by choosing η^0 so that $\mathcal{H}(U^1(\xi^0,\zeta),\eta^0)$ become optimum (thus inequality (2.24) is weakened to the limit). This is (see Appendix 2.1) the random strategy

(2.25) $$\eta_j^0 = \exp\left(-\frac{\alpha}{\mathcal{U}_j^1(\xi^0,\zeta)}\right), \quad (j = 1,\dots,s),$$

where exp(a) is the expression e^{α} and α is the solution of the exponential equation

(2.26) $$\sum_j \exp\left(-\frac{\alpha}{\mathcal{U}_j^1(\xi^0,\zeta)}\right) = 1,$$

the number α being the optimum value of the weighted entropy $\mathcal{H}(U^1(\xi^0,\zeta),\eta^0)$ when η^0 is given by (2.25). The conditions under which there random strategy (2.25) exists providing the correct response, i.e. the weakest conditions under which a game is of type A, will be obtained by replacing (2.25) in (2.22)-(2.24), i.e.

$$\sum_{j} \frac{\alpha}{u_j^1(\xi^0,\zeta)} \exp\left(-\frac{\alpha}{u_j^1(\xi^0,\zeta)}\right) \geq u^2(\xi,\eta,\zeta) + \log\left(\sum_{j} \exp(-u_j^2(\xi^0,\zeta))\right), \tag{2.27}$$

$$\sum_{j} \frac{\alpha}{u_j^1(\xi^0,\zeta)} \exp\left(-\frac{\alpha}{u_j^1(\xi^0,\zeta)}\right) \geq u^3(\xi^0,\eta,\zeta) + \log\left(\sum_{j} \exp(-u_j^3(\xi^0,\zeta))\right), \tag{2.28}$$

$$\alpha \geq \sum_{j} u_j^1(\xi^0,\zeta) - u^1(\xi,\eta,\zeta)\ . \tag{2.29}$$

The relations (2.27), (2.28), (2.29) are called the compatibility relations of the game.

To summarize, a random strategy η^0 assuring correct response must be so chosen that its entropy and weighted entropy will satisfy the inequalities (2.22)-(2.24). If the compatibility relations (2.27)-(2.29) are satisfied, the random strategy η^0 determined effectively by expression (2.25) will ensure the right response, at the same time being the widest random strategy of this kind. (If there is a random strategy ensuring correct response, then the strategy derived from (2.25) will also achieve correct response, though it may happen, in an A type game, that the random strategy given by (2.25) is the only one ensuring this right response!)

We shall now briefly describe an A type game, i. e. a game in which the inequalities (2.22)-(2.24) are satisfied. Consider such a game involving three persons, with independent strategies, in which each player has two pure strategies, i.e.

$r = s = t = 2$.

Let the random strategies of the three players be $\xi = (\xi_1, \xi_2)$, $\eta = (\eta_1, \eta_2)$, $\zeta = (\zeta_1, \zeta_2)$ where

$$\xi_1 = 1/4\,,\quad \xi_2 = 3/4\,,\quad \eta_1 = 3/4\,,\quad \eta_2 = 1/4\,,\quad \zeta_1 = 3/5\,,\quad \zeta_2 = 2/5\,,$$

and the utilities of the game variants for the three players are

$$u^1_{111} = 5,\ u^1_{112} = 5,\ u^1_{121} = 1,\ u^1_{122} = 2,\ u^1_{211} = 3,\ u^1_{212} = 5,\ u^1_{221} = 1,\ u^1_{222} = 1;$$

$$u^2_{111} = 2,\ u^2_{112} = 5,\ u^2_{121} = 7,\ u^2_{122} = 6,\ u^2_{211} = 1,\ u^2_{212} = 1,\ u^2_{221} = 2,\ u^2_{222} = 2;$$

$$u^3_{111} = 3,\ u^3_{112} = 6,\ u^3_{121} = 8,\ u^3_{122} = 5,\ u^3_{211} = 1,\ u^3_{212} = 2,\ u^3_{221} = 1,\ u^3_{222} = 2.$$

The mean payoff of the three players will then be

$$u^1(\xi,\eta,\zeta) = 3{,}350\,;\quad u^2(\xi,\eta,\zeta) = 1{,}950\,;\quad u^3(\xi,\eta,\zeta) = 2{,}2625\,.$$

At a certain moment, player 1 introduces the element of surprise by bringing maximum indetermination into the choice of his own pure strategy, i.e. by passing from random strategy ξ for which uncertainty is $H(\xi) = 0{,}8113$ to random strategy $\xi^0 = (\xi^0_1, \xi^0_2)$ defined by $\xi^0_1 = \xi^0_2 = 1/2$ and having the maximum uncertainty $H(\xi^0) = 1{,}0000$ which also brings him a greater payoff

$$u^1(\xi^0,\eta,\zeta) = 3{,}600\,.$$

Is it possible, in this case, forplayer 2, who changes his random strategy, only to give the right response even though player 3 does not help him in any way? The answer is affirmative, because this game is proved to be an A of type game. Random strategy $\eta^0=(\eta_1^0,\eta_2^0)$ defined by $\eta_1^0=2/3, \eta_2^0=1/3$ and having the entropy $H(\eta^0)=0{,}9180$ and the weighted entropy $\mathfrak{F}(U^1(\xi^0,\zeta),\eta^0)=2{,}3486$ will produce the right response because the inequalities (2.22)-(2.24) are satisfied. Actually, in this case

$$\mathcal{U}_1^2(\xi^0,\zeta)=2{,}10\ ;\quad \mathcal{U}_2^2(\xi^0,\zeta)=4{,}30\ ;$$
$$\mathcal{U}_1^3(\xi^0,\zeta)=2{,}80\ ;\quad \mathcal{U}_2^3(\xi^0,\zeta)=4{,}10\ ;$$

hence inequality (2.24) is verified. Also, in this case $\mathcal{U}_1^1(\xi^0,\zeta)=$ $=4{,}40\ ;\ \mathcal{U}_2^1(\xi^0,\zeta)=1{,}20$ which gives us (2.24), i.e.

$$\sum_{j=1,2}\mathcal{U}_j^1(\xi^0,\zeta)-\mathcal{U}^1(\xi,\eta,\zeta)=5{,}60-3{,}35=2{,}25<\mathfrak{F}(U^1(\xi^0,\zeta),\eta^0)=2{,}3486\,;$$

hence

$$\mathcal{U}^2(\xi,\eta,\zeta)+\log\left(\sum_j 2^{-\mathcal{U}_j^2(\xi^0,\zeta)}\right)=1{,}9500+\log\frac{2^{22/10}+1}{2^{43/10}}<$$

$$<1{,}9500+\log\frac{2^{22/10+1}}{2^{43/10}}=1{,}95-1{,}10=0{,}85<H(\eta^0)=0{,}9180\,.$$

i.e. inequality (2.22) and, similarly,

$$\mathcal{U}^3(\xi,\eta,\zeta) + \log\left(\sum_j 2^{-\mathcal{U}_j^3(\xi^0,\zeta)}\right) = 3{,}125 + \log\frac{2^{13/10}+1}{2^{41/10}} <$$

$$< 3{,}125 + \log\frac{3{,}5}{2^{41/10}} = 3{,}1250 + 1{,}8073 - 4{,}1 = 0{,}8323 < H(\eta^0) = 0{,}9180 .$$

i.e. inequality (2.23), are both satisfied. Consequently, random strategy η^0 provides the right response, i.e. it simultaneously satisfies the inequalities (2.9)-2.11). Actually, a simple calculation gives:

$$\mathcal{U}^1(\xi^0,\eta^0,\zeta) = 3{,}167 < \mathcal{U}^1(\xi,\eta,\zeta) = 3{,}35$$

(the response has made surprise unprofitable to player 1 who introduced it),

$$\mathcal{U}^2(\xi^0,\eta^0,\zeta) = 3{,}767 > \mathcal{U}^2(\xi,\eta,\zeta) = 1{,}950 ,$$

(the mean payoff of player 2 has not decreased),

$$\mathcal{U}^3(\xi^0,\eta^0,\zeta) = 3{,}2333 > \mathcal{U}^3(\xi^0,\eta,\zeta) = 3{,}1250$$

(the mean payoff of player 3 has been affected by the response).

We have now seen what will happen in an A type game, but what happens if the game is not of type A, i.e. if one or more of the compatibility relations are not satisfied? These relations of compatibility will give us the solution in each particular case.

Suppose then, that compatibility relation (2.29) is not satisfied. Random strategy η^0 providing the correct response will exist however, if the third player, renouncing his neutrality and deciding to cooperate with player 2, chooses a random strategy ζ^0 which satisfies compatibility relation (2.29), so that

$$\tilde{\alpha} \geqslant \sum_j \mathcal{U}_j^1(\xi^0,\zeta^0) - \mathcal{U}^1(\xi,\eta,\zeta) \tag{2.30}$$

where $\tilde{\alpha}$ is now the solution of the exponential equation

$$\sum_j \exp\left(-\frac{\tilde{\alpha}}{\mathcal{U}_j^1(\xi^0,\zeta^0)}\right) = 1\,. \tag{2.31}$$

As the solution of such an exponential equation will always be non-negative, if, for instance, player 3 may choose his random strategy ζ^0 such that

$$\sum_j \mathcal{U}_j^1(\xi^0,\zeta^0) \leqslant \mathcal{U}^1(\xi,\eta,\zeta) \tag{2.32}$$

inequality (2.30) will, of course, be automatically satisfied.

Here is a simple case showing that even an inequality of type (2.32) can be realized. Consider a 3-person game, with independent strategies, for which, with the notations used far:

$$\xi_1 = 1/8\,,\ \xi_2 = 7/8\,;\ \eta_1 = 1/4\,,\ \eta_2 = 3/4\,;\ \zeta_1 = 4/5\,,\ \zeta_2 = 1/5\,;\ \ \xi_1^0 = \xi_2^0 = 1/2\,.$$

If the utilities of the game variants for the first player are

$$u^1_{111} = 40\ , \quad u^1_{112} = 5\ , \quad u^1_{121} = 100\ , \quad u^1_{122} = 4\ ;$$

$$u^1_{211} = 10\ , \quad u^1_{212} = 1\ , \quad u^1_{221} = 50\ , \quad u^1_{222} = 2$$

and if player 3 resorts to the new strategy $\zeta^0_1 = 0, \zeta^0_2 = 1$ we see that although

$$\sum_{j=1,2} u^1_j(\xi^0,\zeta) = 63{,}2 < u^1(\xi,\eta,\zeta) = 24{,}4$$

after adoption of strategy ζ^0, we have

$$\sum_{j=1,2} u^1_j(\xi^0,\zeta^0) = 1{,}2 < u^1(\xi,\eta,\zeta) = 24{,}4\ .$$

Consequently, if the inequality (2.29) is not satisfied, it can be corrected through the intervention of player 3 who will choose a random strategy ζ^0 for which we have inequality (2.30). Then, if inequality (2.27), or both, are not satisfied, it becomes necessary to introduce new utilities u^{*3}_{ijk} (of player 3) and u^{*2}_{ijk} (of player 2) to correct them.

Suppose that inequality (2.27) is not satisfied. For simplification we note

$$(2.33) \qquad A(\xi^0,\zeta^0) = \sum_j \frac{\alpha}{u^1_j(\xi^0,\zeta^0)} \exp\left(-\frac{\alpha}{u^1_j(\xi^0,\zeta^0)}\right)\ .$$

It is sufficient to choose u^{*2}_{ijk} so that

$$\mathcal{U}_j^{*2}(\xi^0,\zeta) = \sum_{i,k} u_{ijk}^{*2}\xi_i^0\zeta_k^0 \geq \mathcal{U}^2(\xi,\eta,\zeta) + \log s - A(\xi^0,\zeta) \tag{2.34}$$

for every $j = 1,\ldots,s$ and we have

$$\exp(-\mathcal{U}_j^{*2}(\xi^0,\zeta^0)) \leq \frac{1}{s}\exp(A(\xi^0,\zeta^0) - \mathcal{U}^2(\xi,\eta,\zeta)), \tag{2.35}$$

for every j, i.e.

$$\sum_j \exp(-\mathcal{U}_j^{*2}(\xi^0,\zeta^0)) \leq \exp(A(\xi^0,\zeta^0) - \mathcal{U}^2(\xi,\eta,\zeta))$$

hence

$$A(\xi^0,\zeta^0) \geq \mathcal{U}^2(\xi,\eta,\zeta) + \log\left(\sum_j \exp(-\mathcal{U}_j^{*2}(\xi^0,\zeta^0))\right) \tag{2.36}$$

that is, precisely the compatibility condition (2.27) with the new utilities u_{ijk}^{*2}.

Similarly, if condition (2.28) is not satisfied, it is sufficient to choose the utilities u_{ijk}^{*3} such that to have

$$\sum_{i,k} u_{ijk}^{*3}\xi_i^0\zeta_k^0 \geq \mathcal{U}^3(\xi^0,\eta,\zeta) + \log s - A(\xi^0,\zeta^0) \tag{2.37}$$

that compatibility condition (2.28) be corrected.

The utilities u_{ijk}^{*2} and u_{ijk}^{*3} correcting the compatibility relations (2.27) and (2.28) can always be determined for any random strategies ξ,ξ^0,η,ζ, on the other hand, it may happen that random strategy ζ satisfying (2.30) and correcting (2.29) does not exist.

In short, if the compatibility relations (2.22)-(2.24) are satisfied, the game is of type A and there exists random strategy which provides the right response, which may be chosen as (2.25). If the game is not of type A , then we have seen that compatibility relation (2.29) can be corrected (though not always) by a new strategy ζ^0 of player 3 (which is so chosen as to have (2.30), i.e. practically such that $\sum_j u_j^1(\xi^0, \zeta^0)$ becomes the lowest possible), compatibility condition (2.27) can always be corrected by chosing new utilities u_{ijk}^{*2} (which satisfy, for instance, (2.34), and compatibility condition (2.28) can always be corrected by choosing new utilities u_{ijk}^{*3} (which satisfy, for instance, (2.37), and thus the game becomes an A type game.

Appendix 2.1

LEMMA 1. Let $\eta = (\eta_1, \dots, \eta_s)$ be a probability distribution. Regardless of the numbers $u_j \geq 0$ we have the following inequality

$$\sum_j u_j \eta_j \geq H(\eta) - \log \Phi(u) , \tag{A}$$

where

$$\Phi(u) = \sum_j e^{-u_j} .$$

<u>Proof</u>: We use the evident inequality

$$-\sum_i p_i \log p_i \leq -\sum_i p_i \log q_i ,$$

for every probability distribution

$$p_i \geq 0 , \quad q_i \geq 0 , \quad \sum_i p_i = \sum_i q_i = 1 .$$

Hence

$$H(\eta) = -\sum_j \eta_j \log \eta_j \leq \sum_j \eta_j \log\left(\frac{e^{-u_j}}{\Phi(u)}\right) =$$

$$= \log \Phi(u) + \sum_j e^{-u_j} .$$

where $\quad \Phi(u) = \sum_j e^{-u_j} \qquad$ q.e.d.

LEMMA 2: Let $\eta = (\eta_1, \ldots, \eta_s)$ be a probability distribution. Whatever the real numbers $u_j \geq 0$ we have the inequality

(B) $$\sum_j u_j \eta_j \leq \sum_j u_j - \mathcal{J}(u,\eta),$$

where $u = (u_1, \ldots, u_s)$ and

$$\mathcal{J}(u,\eta) = -\sum_j u_j \eta_j \log \eta_j .$$

Proof: We have the immediate inequality: $\log x < x - 1$, for every $x \neq 1$ and $\log x = x - 1$ if and only if $x = 1$. Then

$$\sum_j u_j \eta_j \log \frac{1}{\eta_j} \leq \sum_j u_j \eta_j \left(\frac{1}{\eta_j} - 1\right) = \sum_j u_j - \sum_j u_j \eta_j$$

whence (B) will result immediately. q.e.d.

LEMMA 3: The weighted entropy

(A1) $$\mathcal{J} = \mathcal{J}(u,\eta) = -\sum_j u_j \eta_j \log \eta_j$$

reaches the optimum α for

$$\eta_j = e^{-\frac{\alpha}{u_j}}, \quad (j = 1, \ldots, s),$$

where α is the solution of the exponential equation

$$\sum_j e^{-\frac{\alpha}{u_j}} = 1 .$$

Proof: We determine the random distribution $(\eta_j)_{1 \leq j \leq s}$ for which (A1) is optimum, compatible with the condition

$$1 = \sum_j \eta_j \,. \tag{A2}$$

Using the method of Lagrangeian multipliers we have

$$\gamma + \sum_j u_j \eta_j - \alpha = -\sum_j u_j \eta_j \log\left[\eta_j \exp\left(\frac{\alpha}{u_j} - 1\right)\right] \leq$$

$$\leq \sum_j u_j \exp\left(-\frac{\alpha}{u_j}\right), \quad (-x\log x \leq e^{-1}),$$

the equality holding if and only if

$$\eta_j = e^{-\frac{\alpha}{u_j}}$$

in which case $\gamma = \alpha$, α being obtained from relation (A2) i.e.

$$\sum_j e^{-\frac{\alpha}{u_j}} = 1 \,. \tag{A3}$$

Remark: The optimum α of the weighted entropy $\gamma(u,\eta)$ given in Lemma 3 corresponds to the random distribution

$$\eta_j = e^{-\frac{\alpha}{u_j}}$$

for which $\gamma(u,\eta) + \sum_j u_j \eta_j$ is maximum.

Appendix 2.2

Using the notation given in this paper we shall prove that if the entropy and weighted entropy of random strategy η^0 satisfy the inequalities (2.22)-(2.24), this strategy will produce the right response, i.e. there occur the inequalities (2.9)-(2.11).

Let us start conversely, from the inequalities (2.9)-(2.11), more exactly from the equivalent inequalities (2.9')-(2.10')-(2.11'). On the basis of inequality (A) in Appendix 2.1, we have

(A4) $$\sum_{i,j,k} u^2_{ijk}\xi^0_i\eta^0_j\zeta_k \geq H(\eta^0) - \log \Phi(u^2,\xi^0,\zeta),$$

where u^2 is the vector of components u^2_{ijk} and where

(A5) $$\Phi(u^2,\xi^0,\zeta) = \sum_j \exp\left(-\sum_{i,k} u^2_{ijk}\xi^0_i\zeta_k\right).$$

Consequently, inequality (2.10') will be satisfied if

(A6) $$H(\eta^0) \geq \mathcal{U}^2(\xi,\eta,\zeta) + \log\left(\sum_j e^{-\mathcal{U}^2_j(\xi^0,\zeta)}\right).$$

Similarly, by again applying inequality (A) from

Appendix 2.1, we shall find that ineqaulity (2.11') is satisfied if

$$H(\eta^0) \geq \mathcal{U}^3(\xi^0,\eta,\zeta) + \log\Big(\sum_j e^{-\mathcal{U}_j^3(\xi^0,\zeta)}\Big). \tag{A7}$$

Applying inequality (B) as given in Appendix 2.1, it follows that

$$\sum_{i,j,k} u^1_{ijk}\,\xi^0_i\,\eta^0_j\,\zeta_k \leq \sum_{i,j,k} u^1_{ijk}\,\xi^0_i\,\zeta_k + \sum_j\Big(\sum_{i,k} u^1_{ijk}\,\xi^0_i\,\zeta_k\Big)\eta_j \log\eta_j .$$

Hence, inequality (2.9') will be satisfied if

$$\begin{aligned}\sum_{i,j,k} u^1_{ijk}\,\xi^0_i\,\zeta_k + \sum_j\Big(\sum_{i,k} u^1_{ijk}\,\xi^0_i\,\zeta_k\Big)\eta_j \log\eta_j &< \\ < \sum_{i,j,k} u^1_{ijk}\,\xi_i\,\eta_j\,\zeta_k &,\end{aligned} \tag{A8}$$

i.e. if

$$\mho(U^1(\xi^0,\zeta),\eta^0) > \sum_j \mathcal{U}^1_j(\xi^0,\zeta) - \mathcal{U}^1(\xi,\eta,\zeta)$$

where $U^1(\xi^0,\zeta)$ is the vector of components

$$U^1(\xi^0,\zeta) = (\mathcal{U}^1_1(\xi^0,\zeta),\ldots,\mathcal{U}^1_s(\xi^0,\zeta)) .$$

Obviously, the inequalities (A6), (A7), (A8) are the same as the inequalities (2.22), (2.23) and (2.24), q.e.d.

Appendix 3

ON THE MOST RATIONAL ALGORITHM OF RECOGNITION

1. Problems of recognition occur in practically every field of human activity as for exemple in medical diagnosis, chemical analysis, recognize of a failure in a complicated mechanism, classification problems, etc. The problem of recognition dealt with in this paper can be described by the following simple model, similar to the RENYI'S model for the theory of random search. Let

$$E_n = \{x_1, x_2, \ldots, x_n\}$$

be a finite set having $n \geq 2$ distinguishable elements - called entities - and suppose that we want to recognize an unknown entity x of the set E_n. The set E_n itself is supposed to be known to us. Let us suppose further that it is not possible to observe the entity x directly, but we may choose some functions from a given set F_N of functions defined on E_n,

$$F_N = \{f_1, f_2, \ldots, f_N\}$$

- called the set of the characteristics of the entities from the set E_n - and observe the values $f_1(x), f_2(x), \ldots, f_N(x)$ taken on by these functions at the unknown entity x. We suppose that the number of different values taken on by every function f belonging to the set F_N , is much smaller than n . Let

$$V_{r_k} = \{f_k^{(1)}, f_k^{(2)}, \ldots, f_k^{(r_k)}\}$$

be the set of the values taken on by the characteristic $f_k \in F_N$. For many particular problems we are especially interested in the case when each characteristic $f \in F_N$ takes on only two values. When n is a large number it is necessary, of course, to observe the value of a large number of characteristics $f \in F_N$ at the entity x . Each such observation gives us only partial information on the entity x , namely it specifies a subset of E_n to which x must belong. But, after making a fairly large number of such observations the information obtained enables us to recognize x . However, we want to recognize x by a not too large number of observations.

LANDA proposed such a strategy of recognition giving the so called most rational algorithm of recognition and applying it to the problem of recognition of sentences in the Russian syntax. According to this algorithm, it is necessary at every moment to choose and to observe firstly such a characteristic f from the set F_N , supplying the largest amount of information, i.e. eliminating the largest degree of uncertainty. Nevertheless, the most rational algorithm of recognition, in the form mentioned above, neglects one very important fact. As a matter-of-fact, we want to recognize the entity x by a not too large number of observations, but in the same time, by a not too large cost. Indeed, we may e.g. suppose that each observation is connected with a certain cost and we want to keep the cost of the whole procedure of recognition relatively low. To verify for example the characteristic f_k having the value $f_k(x) = f_k^{(i)}$ may be more expensive than to verify the same characteristic f_k when it has another value $f_k(x) = f_k^{(j)}$, and then, of course, to verify the characteristic f_k when it has the value $f_k^{(j)}$ is more useful from the point of view of the cost than to verify the same characteristic when it has the value $f_k^{(i)}$.

According to these facts, the reinforcement of the most rational algorithm of recognition needs a measure of information which takes account both aspects of the information, the quantitative and the qualitative one. But in the paper BELIS and GUIASU (1968) such a formula of the information was proposed taking account of the two basic concepts of probability and utility in respect to a goal of all possible events.

Let $\omega_1, \ldots, \omega_n$ be a finite number of events and let $p_1, \ldots, p_n$ be the probabilities of occurrence of these events satisfying to

$$p_i \geqslant 0 \quad (i = 1, \ldots, n); \quad \sum_{i=1}^{n} p_i = 1 .$$

We suppose that the different events $\omega_1, \ldots, \omega_n$ are more or less relevant depending upon the goal to be reached, that is they have different utilities. Let $u_1, \ldots, u_n$ be the utilities of the events $\omega_1, \ldots, \omega_n$, i.e. nonnegative real numbers. The amount of information supplied by an experiment having the events $\omega_1, \ldots, \omega_n$ is

$$\mathcal{H} = \mathcal{H}(u_1,\ldots,u_n,p_1,\ldots,p_n) = -\sum_{i=1}^{n} u_i p_i \log p_i \,. \qquad (3.1)$$

A discussion and an axiomatic treatement of this last formula was given in the paper mentioned above. Of course, the utility of an event is independent of its objective probability of occurence, for instance an event of small probability can have a great utility likewise an event of great probability can have a utility equal to zero with regard to a given goal.

Let us consider the set of the entities E_n and we suppose that the set F_N is a complete system of characteristics for the given set E_n, i.e. for every entity $x_k \in E_n$ there are the indices

$$i_1^k, i_2^k, \ldots, i_{N_k}^k \qquad (N_k \leq N)$$

such that x_k is completely determined by the values

$$f_{i_1^k}(x_k), f_{i_2^k}(x_k), \ldots, f_{i_{N_k}^k}(x_k),$$

where

$$f_{i_j^k}(x_k) \in V_{r_{i_j^k}} \qquad (j = 1,\ldots,N_k).$$

Then we write

$$x_k = f_{i_1^k}(x_k) \wedge f_{i_2^k}(x_k) \wedge \ldots \wedge f_{i_{N_k}^k}(x_k),$$

where the sign Λ represents the conjunction "and". We denote also the conjunction "or" by the sign V. Then it is not difficult to introduce all the characteristics $f_1,\ldots,f_N$. Indeed, if for example we have

$$x_k = f_1(x_k)\Lambda f_2(x_k)\Lambda\ldots\Lambda f_{N-1}(x_k),$$

then the explicit expression of the entity x_k will be

$$\begin{aligned} x_k = {} & [f_1(x_k)\Lambda f_2(x_k)\Lambda\ldots\Lambda f_{N-1}(x_k)\Lambda f_N^{(1)}]V \\ & V[f_1(x_k)\Lambda f_2(x_k)\Lambda\ldots\Lambda f_{N-1}(x_k)\Lambda f_N^{(2)}]V\ldots \\ & \ldots V[f_1(x_k)\Lambda f_2(x_k)\Lambda\ldots\Lambda f_{N-1}(x_k)\Lambda f_N^{(r_N)}], \end{aligned}$$

because always

$$f_k^{(1)}Vf_k^{(2)}V\ldots Vf_k^{(r_k)} = \Omega \qquad (k = 1,\ldots,N)$$

where Ω is the total event, i.e. the sure event. Obviously it is possible that some combinations of some values of the characteristics are not possible (relations of incompatibility). For example it is possible that

$$f_1^{(2)}\Lambda f_2^{(1)}\Lambda f_5^{(3)}\Lambda f_N^{(r_N)} = \Phi,$$

where Φ is the impossible event. Of course, whichever be the possible value $f_k^{(j)}$ we have

$$f_k^{(j)}\Lambda\Phi = \Phi, \quad f_k^{(j)}V\Omega = \Omega$$

$$f_k^{(j)} V \Phi = f_k^{(j)}, \quad f_k^{(j)} \Lambda \Omega = f_k^{(j)}.$$

Let now

$$p(x_k) \geqslant 0 \quad (k = 1, \ldots, n); \quad \sum_{k=1}^{n} p(x_k) = 1$$

be the probabilities of all the possible entities. If we do not know anything about this a priori distribution of the entities we will suppose all these entities $x_1, x_2, \ldots, x_n$ with the same prob̲ability, i.e. $1/n$.

If we have written the explicit expression of all the entities $x_1, x_2, \ldots, x_n$ taking account the incompatibility relations, then it is very easy to count the probabilities of all the values of the characteristics. We denote by

$$p_{jk} = p(f_k^{(j)})$$

the probability of the value $f_k^{(j)}$ of the characteristic f_k .

Let now u_{jk} be the utility in respect to a goal of the same value $f_k^{(j)}$ of the characteristic f_k . Then, according to the expression (3.1), the amount of the information supplied by the observation of the characteristic f_k will be

$$\mathcal{H}_k = \mathcal{H}(f_k) = -\sum_{j=1}^{r_k} u_{jk}\, p_{jk} \log p_{jk} \quad (k = 1, \ldots, N).$$

By the most rational algorithm of recognition it is necessary to choose and to observe firstly the characteristic f_{k_0} so that

$$\mathcal{H}_{k_0} = \max_{1 \leq k \leq N} \mathcal{H}_k .$$

The observation of this characteristic specifies a subset $E'_m \subset E_n (m < n)$, to which x must belong and we will repeat the procedure described above.

Clearly, one problem arises vey naturally, i.e. to establish what are the probabilities of the values of an arbitrary characteristic from the set F_N, supplying the largest amount of the information compatible with the given utilities of these values. This problem is solved by the following theorem.

THEOREM. Let $p_{jk} (j = 1, \ldots, r_k)$, a complete system of probabilities for every $k = 1, \ldots, N$, i.e.

$$p_{jk} \geq 0 \quad (j = 1, \ldots, r_k),$$

(3.2)

$$\sum_{j=1}^{r_k} p_{jk} = 1 ,$$

whichever be $k = 1, \ldots, N$, and let be some utilities, i.e. nonnegative real numbers

$$u_{jk} \geq 0 \quad (j = 1, \ldots, r_k),$$

whichever be $k=1,\ldots,N$. The amount of the information

$$\mathcal{H}_k = -\sum_{j=1}^{r_k} u_{jk} p_{jk} \log p_{jk} \tag{3.3}$$

is maximum if and only if we have

$$p_{jk} = 2^{-\frac{\alpha_k}{u_{jk}}} e^{-1} \; (j = 1,\ldots,r_k), \tag{3.4}$$

for every $k=1,\ldots,N$, where α_k satisfies the equality

$$\sum_{j=1}^{r_k} 2^{-\frac{\alpha_k}{u_{jk}}} = e .$$

The solution α_k of this last equation is just the maximum value of the information $\mathcal{H}_k$

PROOF. It is necessary to find the system of prob̲abilities $p_{jk}(j=1,\ldots,r_k)$, giving the maximum value of the expression (3.3) compatible with the utilities $u_{jk}(j=1,\ldots,r_k)$. Obviously, we have

$$\log x < (x-1)\log e$$

for every $x \neq 1$ and

$$\log x = (x-1)\log e$$

if and only if $x=1$. By using the classical LAGRANGE's multipliers method we obtain

$$\mathcal{H}_k - \alpha_k = \sum_{j=1}^{r_k} u_{jk} p_{jk} \log \frac{1}{p_{jk}} - \alpha_k \sum_{j=1}^{r_k} p_{jk}$$

$$= \sum_{j=1}^{r_k} p_{jk} \left(u_{jk} \log \frac{1}{p_{jk}} - \alpha_k \right)$$

$$= \sum_{j=1}^{r_k} p_{jk} \log \left(\frac{1}{(p_{jk})^{u_{jk}}} 2^{-\alpha_k} \right)$$

$$\leqslant \sum_{j=1}^{r_k} u_{jk} \, 2^{-\frac{\alpha_k}{u_{jk}}} e^{-1} \log e = b_k .$$

The equality holds if and only if

$$p_{jk} = 2^{-\frac{\alpha_k}{u_{jk}}} e^{-1} \quad (j = 1, \ldots, N) , \tag{2.5}$$

and then

$$\mathcal{H}_k = \alpha_k + b_k .$$

Of course, the probabilities (3.5) must verify the condition (3.2), i.e.

$$\sum_{j=1}^{r_k} 2^{-\frac{\alpha_k}{u_{jk}}} = e .$$

REMARK. If we have

$$u_{jk} = u_k \quad (j = 1,\ldots,r_k;\ k = 1,\ldots,N),$$

then, from the theorem mentioned above it results the very well known fact that the entropy

$$\mathcal{H}_k = -u_k \sum_{j=1}^{r_k} p_{jk} \log p_{jk}$$

is maximum if and only if

$$p_{jk} = \frac{1}{r_k} \quad (j = 1,\ldots,r_k),$$

for every $k = 1,\ldots,N$.

Using the theorem proved above, it its possible to estimate what is the characteristic supplying the largest amount of the information compatible with the respective utilities, without to compute exactly the entropies (3.3) of all the possible characteristics.

2. For the whole theory presented above it is very usefully to give an example. Thus we suppose that we have five possible entities

$$E_5 = \{x_1, x_2, x_3, x_4, x_5\},$$

and four characteristics

$$F_4 = \{f_1, f_2, f_3, f_4\}.$$

We suppose also that each characteristic takes on only two values, namely the characteristic f_k, $1 \leq k \leq 4$, takes on the values $\{ f_k^{(1)}, f_k^{(2)} \}$. Let us suppose that the entities of the set E_5 are given by

$$x_1 = f_1^{(1)} \wedge f_4^{(1)}, \quad x_2 = f_1^{(1)} \wedge f_4^{(2)}, \quad x_3 = f_1^{(2)} \wedge f_3^{(1)},$$

$$x_4 = f_1^{(2)} \wedge f_2^{(1)}, \quad x_5 = f_1^{(2)} \wedge f_2^{(2)} \wedge f_3^{(2)},$$

and let

$$f_3^{(1)} \wedge f_4^{(2)} = \Phi, \quad f_2^{(1)} \wedge f_4^{(2)} = \Phi,$$

$$f_1^{(2)} \wedge f_4^{(2)} = \Phi, \quad f_2^{(1)} \wedge f_3^{(1)} = \Phi,$$

be the relations of incompatibility. Using the BOOLE's classical formalism, the explicit expressions of the entities as functions of the characteristics' values are

$$x_1 = (f_1^{(1)} \wedge f_2^{(2)} \wedge f_3^{(1)} \wedge f_4^{(4)}) \vee (f_1^{(1)} \wedge f_2^{(1)} \wedge f_3^{(2)} \wedge f_4^{(1)}) \vee (f_1^{(1)} \wedge f_2^{(2)} \wedge f_3^{(2)} \wedge f_4^{(1)});$$

$$(3.6) \quad x_2 = f_1^{(1)} \wedge f_2^{(2)} \wedge f_3^{(2)} \wedge f_4^{(2)}; \qquad x_3 = f_1^{(2)} \wedge f_2^{(2)} \wedge f_3^{(1)} \wedge f_4^{(1)};$$

$$x_4 = f_1^{(2)} \wedge f_2^{(1)} \wedge f_3^{(2)} \wedge f_4^{(1)}; \qquad x_5 = f_1^{(2)} \wedge f_2^{(2)} \wedge f_3^{(2)} \wedge f_4^{(1)}.$$

We suppose that all the entities of the set E_5 have the same a priori probabilities, i.e.

$$p(x_1) = p(x_2) = p(x_3) = p(x_4) = p(x_5) = \frac{1}{5}. \qquad (3.7)$$

Let now x be an arbitrary entity and we want to recognize the entity x by a not too large number of observations and in the same time by a not too large cost. Obviously

$$x = x_1 \vee x_2 \vee x_3 \vee x_4 \vee x_5 . \qquad (3.8)$$

According to the equalities (3.6), (3.7) and (3.8) we obtain for the probabilities of the characteristics' values

$$p_{11} = p(f_1^{(1)}) = \frac{2}{5}, \quad p_{21} = p(f_1^{(2)}) = \frac{3}{5},$$
$$p_{12} = p(f_2^{(1)}) = \frac{4}{15}, \quad p_{22} = p(f_2^{(2)}) = \frac{11}{15},$$
$$p_{13} = p(f_3^{(1)}) = \frac{4}{15}, \quad p_{23} = p(f_3^{(2)}) = \frac{11}{15},$$
$$p_{14} = p(f_4^{(1)}) = \frac{4}{5}, \quad p_{24} = p(f_4^{(2)}) = \frac{1}{5}.$$

We suppose now that each observation is connected with a certain utility in respect to a goal. From this point of view we suppose now that we have

$$u_{11} = 1, \quad u_{21} = 9, \quad u_{12} = 7, \quad u_{22} = 3,$$
$$u_{13} = 1, \quad u_{23} = 10, \quad u_{14} = 3, \quad u_{24} = 7, \qquad (3.9)$$

where $u_{jk}(k = 1,2,3,4; j = 1,2)$, represents the utility of the value $f_k^{(j)}$ of the characteristic f_k, in respect to our goal (for example in respect to the smallest cost of the whole procedure). From (3.9), according to the expression (3.3), we obtain for the amount of the information supplied by the observations of the characteristics of the set E_5 :

$$\mathcal{H}_1 = \mathcal{H}(f_1) = 4.5086 \; ; \qquad \mathcal{H}_2 = \mathcal{H}(f_2) = 4.5464 \; ;$$

$$\mathcal{H}_3 = \mathcal{H}(f_3) = 3.7937 \; ; \qquad \mathcal{H}_4 = \mathcal{H}(f_4) = 4.0233 \, .$$

Thus it is necessary to verify first the characteristic f_2 , supplying the largest amount of information compatible with given utilities. If we verify this second characteristic we may obtain only the two possibilities $f_2^{(1)}$ and respectively $f_2^{(2)}$.

a) We suppose now that we have found the value $f_2^{(1)}$. Then, according to the expression (3.6) it results that we have

$$x = x_1 \vee x_4 \, ,$$

with the probabilities

$$p(x_1) = p(x_4) = \frac{1}{2} \, .$$

Taking into account the expressions (3.6) of the entities x_1 and x_4 we obtain

$$p_{11} = \frac{1}{2} \, , \quad p_{21} = \frac{1}{2} \, , \quad p_{12} = 1 \, , \quad p_{22} = 0 \, , \quad p_{13} = 0 \, ,$$

$$p_{23} = 1 \, , \quad p_{14} = 1 \, , \quad p_{24} = 0 \, .$$

Then, we have

$$\mathcal{H}_1 = 5 \; ; \quad \mathcal{H}_2 = \mathcal{H}_3 = \mathcal{H}_4 = 0 \; .$$

Of course, in this case it is necessary to verify the characteristic f_1 and if we obtain the value $f_1^{(1)}$ it results that $x = x_1$, and if we obtain the other value, namely $f_1^{(2)}$, it results that $x = x_4$.

b) Let us suppose that verifying the characteristic f_2 we have found the value $f_2^{(2)}$. Then, according to the expressions (3.6) it results that we may have the possibilities

$$x = x_1 \vee x_2 \vee x_3 \vee x_5 \; , \tag{3.10}$$

with the probabilities

$$p(x_1) = p(x_2) = p(x_3) = p(x_5) = \frac{1}{4} \; . \tag{3.11}$$

From the expressions (3.6) of the entities x_1, x_2, x_3 and x_5 we have

$$\begin{aligned} p_{11} &= \frac{1}{2}, \quad p_{21} = \frac{1}{2}, \quad p_{12} = 0, \quad p_{22} = 1, \\ p_{13} &= \frac{3}{8}, \quad p_{23} = \frac{5}{8}, \quad p_{14} = \frac{3}{4}, \quad p_{24} = \frac{1}{4}. \end{aligned} \tag{3.12}$$

and we obtain from (3.3), (3.9) and (3.12)

$$\mathcal{H}_1 = 5 \; ; \quad \mathcal{H}_2 = 0 \; ; \quad \mathcal{H}_3 = 4.7686 \; ; \quad \mathcal{H}_4 = 5.4339 \; .$$

It results that it is necessary to verify now the characteristic f_4 . If for this characteristic we obtain the value $f_4^{(2)}$ then we have $x = x_2$ and if we obtain the other value $f_4^{(1)}$ we have still three possibilities

$$x = x_1 V x_3 V x_5 ,$$

with the probabilities

$$p(x_1) = p(x_3) = p(x_5) = \frac{1}{3} .$$

In this new situation, using the same techniques as above, we obtain

$$p_{11} = \frac{1}{3}, \quad p_{21} = \frac{2}{3}, \quad p_{12} = 0, \quad p_{22} = 1, \quad p_{13} = \frac{1}{2} .$$

$$p_{23} = \frac{1}{2}, \quad p_{14} = 1, \quad p_{24} = 0,$$

and therefore

$$\mathcal{K}_1 = 4.0273 \; ; \quad \mathcal{K}_2 = 0 \; ; \quad \mathcal{K}_3 = 5.5 \; ; \quad \mathcal{K}_4 = 0 .$$

Verifying the characteristic f_3 we can obtain: b_1) the value $f_3^{(1)}$ and we have

$$x = x_1 V x_3 ,$$

with

$$p(x_1) = p(x_3) = \frac{1}{2} .$$

In this case

$$p_{11} = \frac{1}{2}, \quad p_{21} = \frac{1}{2}, \quad p_{12} = 0, \quad p_{22} = 1, \quad p_{13} = 1,$$

$$p_{23} = 0, \quad p_{14} = 1, \quad p_{24} = 0,$$

and if we obtain for the characteristic f_1 the value $f_1^{(1)}$ we have $x = x_1$, and if we obtain the other value $f_1^{(2)}$ we have $x = x_3$.

b_2) the value $f_3^{(2)}$ and we have

$$x = x_1 V x_5,$$

with

$$p(x_1) = p(x_5) = \frac{1}{2}.$$

In this last situation

$$p_{11} = \frac{1}{2}, \quad p_{21} = \frac{1}{2}, \quad p_{12} = 0, \quad p_{22} = 1, \quad p_{13} = 0,$$

$$p_{23} = 1, \quad p_{14} = 1, \quad p_{24} = 0,$$

and verifying the characteristic f_1 , if we obtain the value $f_1^{(1)}$ we have $x = x_1$, and if we obtain the value $f_1^{(2)}$ we have $x = x_5$.

Synthesizing all these results we obtain the Fig. 1 of the most rational algorithm of recognition compatible with the utilities (3.9) (the number in the circle represents the number of the characteristic which must be verified):

REMARK. If in the same example as above we suppose

that all the values of the characteristics have the same utilities, i.e.

$$u_{11} = u_{21} = u_{12} = u_{22} = u_{13} = u_{23} = u_{14} = u_{24} = 1 ,$$

the most rational algorithm of recognition give us the Fig. 2.

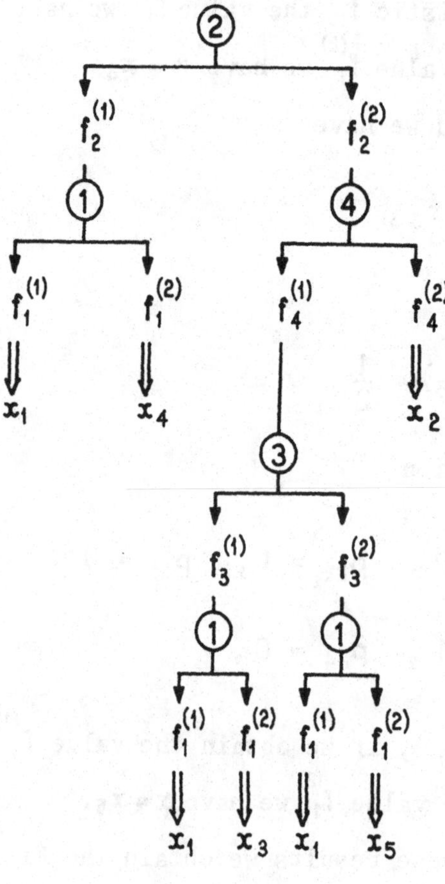

Fig.1

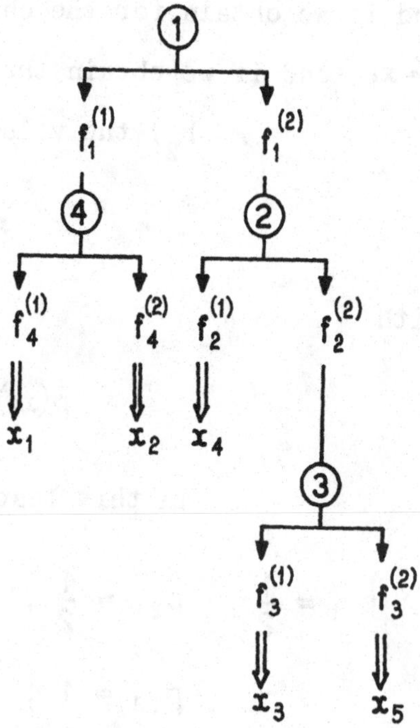

Fig.2

References

[1] Arbib M.A., Give'on Y. Inform. Control 12, 331, 1968.

[2] Arbib M.A., Give'on Y. Inform. Control 12, 346, 1968.

[3] Bacon G.C. IEEE Trans. Circuit Theory CT-11, 307, 1964.

[4] Belis M., Guiasu S., IEEE Trans. Inform. Theory IT-14, 593, 1968.

[5] Billingsley P., Ergodic Theory and Information, Wiley, New York, 1965.

[6] Eilenberg S., Wright J.B., Inform. Control 11, 452, 1967.

[7] Feinstein A., Foundations of Information Theory, McGraw-Hill Book, New York, 1958.

[8] Freyd P., Abelian categories. Harper, New York, 1964.

[9] Guiasu S., Proc. Colloq. Inform. Theory Debrecen (Hungary) 219, 1967.

[10] Guiasu S., Inform. Control, 12, 277, 1968.

[11] Guiasu S., Inform. Control, 16, 103, 1970.

[12] Guiasu S., Applicatii ale teoriei informatiei. Sisteme dinamice. Sisteme cibernetice. Editura Academiei, Bucuresti, 1968.

[13] Guiasu S., Theodorescu R., La théorie mathématique de l'information. Dunod Paris 1968.

[14] Onicescu O., Calcolo delle probabilità ed applicazioni. Veschi, Roma, 1969.

[15] Onicescu O., Guiasu S., Z. Wahrscheinlichkeitstheorie, 1965.

[16] Rabin O., Inform. Control, 6, 230, 1963.

[17] Watanabe S., Knowing and guessing, Wiley New York 1969.

[18] Guiasu S., Malita M., Triade. Editura Stüntifica. Bucuresti, 1973.

[19] Guiasu S., Reports on Mathematical Physics, 2, 165, 1971.

[20] Longo G., Source coding theory, International Centre for Mechanical Sciences, Courses and lectures no. 32, Udine, 1970.

[21] Ingarden R.S., Fortschr. Physik, 12, 567, 1964; 13, 755, 1965.

[22] Ingarden R.S., Urbanik K., Colloq. Math. 9, 281, 1962.

Contents

Printed in the United States
By Bookmasters